Sahil Dhoked
The University of Texas at Dallas
Richardson, TX, USA

Wojciech Golab
University of Waterloo
Waterloo, ON, Canada

Neeraj Mittal
The University of Texas at Dallas
Richardson, TX, USA

ISSN 2155-1626   ISSN 2155-1634 (electronic)
Synthesis Lectures on Distributed Computing Theory
ISBN 978-3-031-20004-5   ISBN 978-3-031-20002-1 (eBook)
https://doi.org/10.1007/978-3-031-20002-1

This Springer imprint is published by the registered company Springer Nature Switzerland AG
The registered company address is: Gewerbestrasse 11, 6330 Cham, Switzerland

Sahil Dhoked · Wojciech Golab · Neeraj Mittal

Recoverable Mutual Exclusion

 Springer

Synthesis Lectures on Distributed Computing Theory

Series Editor

Michel Raynal, Rennes, France

The series publishes short books on topics pertaining to distributed computing theory. The scope largely follows the purview of premier information and computer science conference. Potential topics include, but not are limited to: distributed algorithms and lower bounds, algorithm design methods, formal modeling and verification of distributed algorithms, and concurrent data structures.

Preface

Lock-based synchronization remains one of the most popular techniques in practice for coordinating access to a shared resource by a collection of concurrent processes and can be traced back to Dijkstra's seminal work in the 1960s. While the vast majority of research on this topic has considered reliable processes, the focus has been shifting slowly toward models where processes may fail by crashing, and potentially recover. One of the driving forces behind this trend is the recent emergence of persistent main memory, which allows applications to recover a substantial portion of their state directly after a failure, rather than painstakingly rebuilding it using checkpoints and recovery logs saved in secondary storage.

Inspired by the new possibilities offered by multiprocessor architectures equipped with persistent memory, Golab and Ramaraju recently formalized a fault-tolerant variation of the classic mutual exclusion problem, called Recoverable Mutual Exclusion (RME). Their conceptual model abstracts away many of the low-level technicalities in earlier practitioner-oriented work on fault-tolerant locks, such as how failures are detected and how a stuck critical section is reclaimed, using a simple but powerful assumption: a process that crashes while accessing a recoverable lock must eventually recover and make another attempt to acquire and release the lock.

This monograph describes a growing body of research on the RME problem since its inception in 2016. We have made an attempt to include all the major contributions published to date, and also convey some perspective regarding how the problem itself is evolving. The results are described at a high level to enable readers to quickly understand the main algorithmic ideas. For further technical details and analysis of correctness, the audience is encouraged to read the original research papers, many of which appear in the proceedings of top distributing theory conferences such as PODC and DISC.

The monograph is intended mainly for researchers and serves as a comprehensive introduction to the RME problem. Within it, you will find answers to subtle questions such as what is the precise relationship between RME and persistent memory, why can a process not simply resume its execution from the exact point of failure, how can compositions of locks be recovered correctly, how exactly is the order of entry into the critical section decided in a first-come first-served recoverable lock, and what is the proper way to formalize liveness properties in an environment where dense failures can fundamentally prevent processes from progressing. We also comment extensively on the time and space complexity of different RME algorithms, where the former can be quantified in the same manner as for conventional locks—by counting Remote Memory References (RMRs). As our coverage of known RMR bounds shows, the RME problem is strictly harder to solve in a formal sense than the classic mutual exclusion problem, and does not admit solutions with constant RMR complexity even when the multiprocessor supports certain combinations of powerful read-modify-write instructions. Thus, non-trivial lower bounds can be devised for the RME problem without relying on the somewhat artificial assumption that only reads and writes are supported, which is a recurring theme in theoretical research on the conventional mutual exclusion problem.

The authors welcome your feedback or criticism regarding the content of the monograph and our chosen style of presentation. Please feel free to contact us by email (sahil.dhoked@utdallas.edu, wgolab@uwaterloo.ca, neerajm@utdallas.edu), or approach us directly at conferences. We would be happy to speak to you, and are grateful for your help in improving future editions of the monograph.

Richardson, USA   Sahil Dhoked
Waterloo, Canada   Wojciech Golab
Richardson, USA   Neeraj Mittal
July 2022

Acknowledgments

This work was supported, in part, by the National Science Foundation (NSF) under grant number CNS-1619197, a Google Faculty Research Award, an Ontario Early Researcher Award, and the Natural Sciences and Engineering Research Council of Canada (NSERC).

The authors would like to thank Michel Raynal for inviting them to write this monograph. We are also grateful to Aditya Ramaraju, Danny Hendler, Prasad Jayanti, Anup Joshi, Siddhartha Jayanti, David Yu Cheng Chan, and Philipp Woelfel for allowing us to present their work.

Much of Wojciech Golab's work on this monograph took place on the traditional territory of the Attawandaron, Anishinaabeg, and Haudenosaunee peoples. The main campus of the University of Waterloo is situated on the Haldimand Tract, the land granted to the Six Nations that includes six miles on each side of the Grand River.

January 2022

Sahil Dhoked
Wojciech Golab
Neeraj Mittal

Contents

About the Authors

Sahil Dhoked received his B.Tech. degree in computer science and engineering from the Indian Institute of Technology, Indore in 2013. He completed his Ph.D. degree in computer science from the University of Texas at Dallas in 2022 under the supervision of Dr. Neeraj Mittal. As of March 2023, he is currently employed at Meta Platforms as a research scientist.

Wojciech Golab received his Ph.D. degree in computer science from the University of Toronto in 2010. In the same year, he completed a post-doctoral fellowship at the University of Calgary and later joined Hewlett-Packard Labs in Palo Alto as a Research Scientist. In 2012, he became a faculty member in the Department of Electrical and Computer Engineering at the University of Waterloo. He is broadly interested in concurrency and fault tolerance in distributed systems, with a special focus on bridging the gap between theory and practice. The ACM Computing Reviews recognized one of Professor Golab's papers on shared memory algorithms among 91 others in the "notable computing items published in 2012," and several of his other publications have been distinguished with best paper awards and journal invitations.

Neeraj Mittal received his B.Tech. degree in computer science and engineering from the Indian Institute of Technology, Delhi in 1995 and the M.S. and Ph.D. degrees in computer science from the University of Texas at Austin in 1997 and 2002, respectively. He is currently a professor in the Department of Computer Science at the University of Texas at Dallas and a co-director of the Advanced Networking and Dependable System Laboratory (ANDES). His research interests include multi-core computing, distributed computing, fault-tolerant computing, and distributed algorithms for wireless networking.

Introduction

One of the most commonly used techniques to handle contention in a concurrent system is to use *mutual exclusion (ME)*. The mutual exclusion problem was first defined by Dijkstra more than half a century ago in [1], and later formalized by Lamport [2, 3]. Using locks that provide mutual exclusion enables a process to execute its *critical section* (part of the program that involves accessing shared resources) in isolation without worrying about interference from other processes. Properly used locks mitigate race conditions, thereby ensuring that the system always stays in a consistent state and produces correct outcomes under all scenarios.

Generally, algorithms for mutual exclusion are designed with the assumption that failures do not occur at inopportune times, such as while a process is accessing a lock or a shared resource. However, such failures can occur in the real world. A power outage or network failure might create an unrecoverable situation causing processes to stall or enter an erroneous state. Traditional mutual exclusion algorithms, which are not designed to operate properly in the presence of failures, may fail to guarantee vital correctness properties under such adverse conditions. For example, deadlock could arise if a failure occurs while some process is in the critical section of a lock, leading to potentially disastrous consequences for users of a mission-critical system. This observation gives rise to the *recoverable mutual exclusion (RME) problem*. The RME problem involves designing an algorithm that ensures mutual exclusion, along with other necessary correctness properties, under the assumption that process failures may occur at *any* point during their execution, but the system is able to resurrect failed processes to facilitate recovery.

Traditionally, concurrent algorithms use checkpointing and logging to tolerate failures by regularly saving the relevant portion of application state to a secondary storage device, such as a hard disk drive (HDD) or solid state drive (SSD). Accessing such a device is orders of magnitude slower than accessing main memory. As a result, checkpointing and logging algorithms tend to create performance bottlenecks, even when they are designed to minimize disk accesses. *Persistent memory (PMem)* is a new class of memory technologies

© The Author(s), under exclusive license to Springer Nature Switzerland AG 2023
S. Dhoked et al., *Recoverable Mutual Exclusion*, Synthesis Lectures on
Distributed Computing Theory, https://doi.org/10.1007/978-3-031-20002-1_1

that challenges this limitation by combining the low latency and high bandwidth of traditional random access memory with the density, non-volatility, and economic characteristic of traditional storage media (e.g., hard disk drive). Existing checkpointing and logging algorithms can be modified to use PMem instead of disks to yield better performance, but, in doing so, we would not be leveraging the true power of PMem [4–6]. Notably, PMem can be used to directly store program variables, and in that sense has the potential to enable near-instantaneous recovery from failures.

Most of the application data can be easily recovered after failures by directly storing program variables in PMem. However, recovery of program variables alone is not enough. Processor state information such as the contents of general and special purpose CPU registers (e.g., program counter, condition code register, stack pointer, etc.) as well as contents of cache cannot always be recovered fully. In other words, recovery may be lossy, and, if not handled properly, a failure may cause the system to behave erroneously upon recovery. Due to this reason, there is a renewed interest in developing fast and dependable algorithms, using PMem, for solving many important computing problems in software systems that are vulnerable to process failures.

In this monograph, we describe the recent work that has been published on formulating and solving the RME problem. The RME problem in the current form was formally defined in 2016 by Golab and Ramaraju in [7, 8]. Since then, several research works have been published to address and solve this problem [7–18]. These research works study the problem under a variety of conditions, including but not limited to (i) alternative execution models for use by concurrent applications (e.g., can an application skip acquiring a lock after restarting from a crash if it already held the lock prior to crashing), (ii) whether processes fail independently or simultaneously (individual process failure or system-wide failure), (iii) how a failure impacts the performance of an RME algorithm (non-adaptive, sudden degradation or graceful degradation), and (iv) whether or not a request for the critical section can be aborted. We restrict our attention to solutions that use memory access instructions supported on current hardware platforms. Thus, RME algorithms such as those in [12, 19] that use fetch-and-store-and-store (FASAS), a read-modify-write instruction that is currently not supported on any hardware to the best of our knowledge, have been omitted.

One of the most important measures of performance of a shared memory algorithm, including an RME algorithm, is the maximum number of *remote memory references (RMRs)* made by a process per critical section request in order to acquire and release the lock as well as recover the lock after a failure. Intuitively, RMR complexity captures the number of "expensive" steps performed by a process. Whether or not a memory reference is considered an RMR depends on the underlying memory model. The two most common memory models used to analyze the performance of an RME algorithm are *cache-coherent (CC)* and *distributed shared memory (DSM)* models. Roughly speaking, a step is considered to incur an RMR in the CC model if it causes a memory location to be cached locally or a remote cached copy to be invalidated, and in the DSM model if it accesses data stored on a remote memory module. The CC model captures the working of the caching system used by hardware manufacturers to mask the high latency of memory, whereas the DSM model captures the NUMA (non-uniform memory access) effect observed in large multiprocessors when memory is partitioned into multiple modules.

RMR complexity has been a topic of growing importance in the wake of the multi-core revolution. Building on a series of earlier results [20–23], Attiya, Hendler, and Woelfel [24] proved that any (traditional) ME algorithm that uses only read, write and comparison-based read-modify-write instructions must incur Ω $(\log n)$ RMRs per passage in the worst case. Naturally, the same lower bound also applies to the RME problem. Recently, Chan and Woelfel have proved a lower bound of Ω $(\log n / \log \log n)$ on the worst-case RMR complexity of any RME algorithm that uses a broader set of hardware instructions, including non-comparison based instructions such as fetch-and-store and fetch-and-add, and practical word size of $\Theta(\log n)$ [25]. In contrast, many ME algorithms with worst-case RMR complexity of O (1) are known in the literature for the CC model [26–28] as well as the DSM model [27, 28] based on the same set of instructions. This observation underscores the inherent algorithmic difficulty of the RME problem, and the corresponding opportunity for advancement of scientific knowledge in the field of distributed computing theory.

The rest of the monograph is organized as follows. Chapter 2 provides a brief overview of persistent memory, the main focus of this monograph. Chapter 3 describes the prior work in the area of designing fault tolerant mutual exclusion algorithms, not necessarily limited to using persistent memory. Chapter 4 provides a formal description of the RME problem, including different execution models along with correctness and other desirable properties of RME algorithms. RME algorithms that use only read and write instructions, making them highly portable, are described in Chap. 5. Chapter 6 describes RME algorithms with (optimal) sub-logarithmic RMR complexity. Chapter 7 describes adaptive RME algorithms whose RMR complexity depends on the number of failures that occurred in the system. Chapter 8 describes an RME algorithm with constant amortized RMR complexity. Chapter 9 describes a extension of the RME problem, referred to as *abortable* RME problem, in which a process may withdraw its intention to execute its critical section. Chapter 11 describes RME algorithms specifically designed to handle system-wide failures, when all processes crash at the same time. Finally, Chap. 12 outlines some problems for future research.

To make it easier to compare different RME algorithms, some of the algorithms have been modified from their original versions to use the execution model of an RME lock as proposed by Golab and Ramaraju (GR) in [7, 8]. This is because of the following reasons. First, most RME algorithms in the literature assume the GR model of execution. Second, it is simpler to use since it allows fewer execution paths. Third, it is more portable since it does not require using goto statements between procedures, which are not permitted in most high level programming languages including C, C++ and Java. Also, the monograph does not include proofs of correctness and complexity analysis; the reader is encouraged to consult the original research publications for the omitted material.

References

1. Edsger W. Dijkstra. Solution of a problem in concurrent programming control. *Communications of the ACM (CACM)*, 8(9):569, 1965.
2. Leslie Lamport. The mutual exclusion problem: part I – a theory of interprocess communication. *Journal of the ACM (JACM)*, 33(2):313–326, 1986.
3. Leslie Lamport. The mutual exclusion problem: part II – statement and solutions. *Journal of the ACM (JACM)*, 33(2):327–348, 1986.
4. Dushyanth Narayanan and Orion Hodson. Whole-system persistence. In *Proc. of the 17th International Conference on Architectural Support for Programming Languages and Operating Systems (ASPLOS)*, pages 401–410, New York, NY, USA, March 2012. ACM.
5. Andy Rudoff. Programming models for emerging non-volatile memory technologies. *login Usenix Magazine*, 38(3), 2013.
6. Andy Rudoff. Persistent memory programming. *login Usenix Magazine*, 42(2), 2017.
7. Wojciech Golab and Aditya Ramaraju. Recoverable mutual exclusion. In *Proc. of the 35th ACM Symposium on Principles of Distributed Computing (PODC)*, pages 65–74, 2016.
8. Wojciech Golab and Aditya Ramaraju. Recoverable mutual exclusion. *Distributed Computing (DC)*, 32(6):535–564, 2019.
9. Wojciech Golab and Danny Hendler. Recoverable mutual exclusion in sub-logarithmic time. In *Proc. of the 36th ACM Symposium on Principles of Distributed Computing (PODC)*, pages 211–220, 2017.
10. Wojciech Golab and Danny Hendler. Recoverable mutual exclusion under system-wide failures. In *Proc. of the 37th ACM Symposium on Principles of Distributed Computing (PODC)*, pages 17–26, 2018.
11. Prasad Jayanti and Anup Joshi. Recoverable FCFS mutual exclusion with wait-free recovery. In *Proc. of the 31th International Symposium on Distributed Computing (DISC)*, pages 30:1–30:15, 2017.
12. Prasad Jayanti, Siddhartha V. Jayanti, and Anup Joshi. Optimal recoverable mutual exclusion using only FASAS. In *Proc. of 6th International Conference on Networked Systems (NETYS)*, pages 191–206, 2018.
13. Prasad Jayanti and Anup Joshi. Recoverable mutual exclusion with abortability. In *Proc. of 7th International Conference on Networked Systems (NETYS)*, pages 217–232, 2019.
14. Prasad Jayanti and Anup Joshi. Recoverable mutual exclusion with abortability. arXiv:2012.03140v1, 2020.
15. Prasad Jayanti, Siddhartha Jayanti, and Anup Joshi. A recoverable mutex algorithm with sub-logarithmic RMR on both CC and DSM. In *Proc. of the 38th ACM Symposium on Principles of Distributed Computing (PODC)*, pages 177–186, 2019.
16. Sahil Dhoked and Neeraj Mittal. An adaptive approach to recoverable mutual exclusion. In *Proc. of the 39th ACM Symposium on Principles of Distributed Computing (PODC)*, PODC '20, pages 1–10, New York, NY, USA, 2020. Association for Computing Machinery.

17. David Yu Cheng Chan and Philipp Woelfel. Recoverable mutual exclusion with constant amortized RMR complexity from standard primitives. In *Proc. of the 39th ACM Symposium on Principles of Distributed Computing (PODC)*, New York, NY, USA, August 2020.

18. Daniel Katzan and Adam Morrison. Recoverable, abortable, and adaptive mutual exclusion with sublogarithmic RMR complexity. In *Proc. of the 24th International Conference on Principles of Distributed Systems (OPODIS)*, pages 15:1–15:16, 2021.

19. Aditya Ramaraju. RGLock: Recoverable mutual exclusion for non-volatile main memory systems. Master's thesis, University of Waterloo, 2015.

20. James H. Anderson and Yong-Jik Kim. An improved lower bound for the time complexity of mutual exclusion. *Distributed Computing (DC)*, 15(4):221–253, 2002.

21. Robert Cypher. The communication requirements of mutual exclusion. In *Proc. of the 7th ACM Symposium on Parallel Algorithms and Architectures (SPAA)*, pages 147–156, 1995.

22. Rui Fan and Nancy Lynch. An $\Omega(n \log n)$ lower bound on the cost of mutual exclusion. In *Proc. of the 25th ACM Symposium on Principles of Distributed Computing (PODC)*, pages 275–284, 2006.

23. Wojciech Golab, Vassos Hadzilacos, Danny Hendler, and Philipp Woelfel. RMR-efficient implementations of comparison primitives using read and write operations. *Distributed Computing (DC)*, 25(2):109–162, 2012.

24. Hagit Attiya, Danny Hendler, and Philipp Woelfel. Tight RMR lower bounds for mutual exclusion and other problems. In *Proc. of the 40th ACM Symposium on Theory of Computing (STOC)*, pages 217–226, 2008.

25. David Yu Cheng Chan and Philipp Woelfel. A tight lower bound for the RMR complexity of recoverable mutual exclusion. In *Proc. of the 40th ACM Symposium on Principles of Distributed Computing (PODC)*, 2021.

26. Thomas E. Anderson. The performance of spin lock alternatives for shared-memory multiprocessors. *IEEE Transactions on Parallel and Distributed Systems (TPDS)*, 1(1):6–16, 1990.

27. John M. Mellor-Crummey and Michael L. Scott. Algorithms for scalable synchronization on shared-memory multiprocessors. *ACM Transactions on Computer Systems (TOCS)*, 9(1):21–65, 1991.

28. R. Dvir and G. Taubenfeld. Mutual exclusion algorithms with constant RMR complexity and wait-free exit code. In James Aspnes, Alysson Bessani, Pascal Felber, and João Leitão, editors, *Proc. of the International Conference on Principles of Distributed Systems (OPODIS)*, volume 95, pages 17:1–17:16, Dagstuhl, Germany, October 2017. Schloss Dagstuhl–Leibniz-Zentrum fuer Informatik.

Persistent Memory

The RME problem is inspired by various efforts over the last decade to commercialize persistent memory (PMem) devices that combine the latency benefits of conventional DRAM-based main memories with the non-volatility and high data density of secondary storage devices such as hard disk drives (HDDs) and solid state drives (SDDs). In 2019, these efforts bore long-awaited fruit as Intel released the Optane Persistent Memory alongside the second generation of the Xeon Scalable Processor, which supports specialized persistence instructions for flushing data from the volatile cache to PMem.[1] This revolutionary technology and its programming model [1, 2] allow program state, including low-level program variables and more complex in-memory data structures, to survive system-wide failures caused by power outages without relying on a redundant power supply that may itself be prone to failure.

Figure 2.1 illustrates Intel's implementation of PMem across two successive generations of the Xeon Scalable Processor. On a second generation platform (Fig. 2.1a), which was the first to support Optane persistent memory, PMem resides at the same layer of the memory hierarchy as DRAM, and all layers above main memory are volatile. Persistence instructions such as cache line flushes and write-backs simply push data from the volatile cache to the memory controller's *write pending queue (WPQ)*. During a power failure, any data that remains in the WPQ is saved to Optane memory modules using the residual energy accumulated in the multiprocessor's power supply. Thus, the power fail protected domain includes the WPQ but excludes the cache and CPU registers. In particular, no attempt is made to save the program counter or stack pointer, because both the executable code and call stack of a process are maintained in volatile DRAM (and similarly for the operating system itself). On a third generation Xeon Scalable Processor (Fig. 2.1b), the power fail

---

[1] More precisely, the persistence instructions force data into the processor's persistence domain, which includes certain components of the integrated persistent memory controller.

© The Author(s), under exclusive license to Springer Nature Switzerland AG 2023   7
S. Dhoked et al., *Recoverable Mutual Exclusion*, Synthesis Lectures on
Distributed Computing Theory, https://doi.org/10.1007/978-3-031-20002-1_2

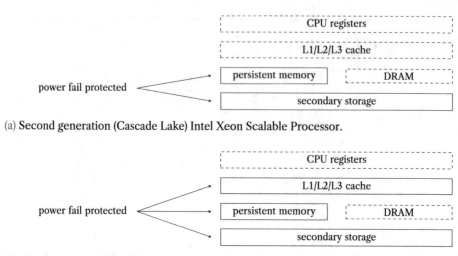

(a) Second generation (Cascade Lake) Intel Xeon Scalable Processor.

(b) Third generation (Ice Lake) Intel Xeon Scalable Processor.

Fig. 2.1 The memory hierarchy and power fail protected domains on Intel's 2nd and 3rd generation Xeon Scalable Processors

protected domain is extended to include the cache, and additional energy from a special battery is used to flush the cache during a power failure. This allows applications to forego cache line flushes and write-backs,[2] and avoids the need to wait for such instructions to complete.

The relationship between the RME problem and persistent memory can be summarized as follows: RME algorithms can be designed to tolerate a variety of failure modes, some (but not all) of which are served well by the unique properties of persistent memory. The specific failure mode that connects RME directly to PMem is an unexpected system reboot (cold boot), such as due to a power failure, which destroys the portion of a program's state that resides in volatile layers of the memory hierarchy. In the literature on the RME problem (e.g., [3]), this mode is known as a system-wide failure, which represents the simultaneous crash of all processes in the system. System-wide failures are considered less severe than individual process failures in the sense that any RME algorithm designed to tolerate the latter automatically tolerates the former. Intuitively, this observation follows because a system-wide failure can be modelled as the correlated occurrence of individual process failures.

The RME problem remains nontrivial even in the presence of persistent memory for the simple reason that program state is spread over multiple levels of the memory hierarchy, some of which continue to be implemented using volatile media. As explained earlier in reference to Intel hardware, these volatile layers generally include CPU registers, which the processor uses to return the responses of memory operations, and whose implementation in hardware is constrained by stringent access latency requirements. Moreover, modern

---

[2] Store fences are still required.

operating systems maintain a program's call stack in volatile main memory (DRAM), even on platforms equipped with PMem. Thus, persistent memory only protects a certain portion of a program's state, and application-specific recovery logic is necessary to correctly revive a collection of processes following a system-wide failure that erases data held in volatile media.

The observant reader will note that our coverage of RME algorithms in the remainder of this monograph excludes any discussion of persistence instructions. This is done intentionally to simplify presentation, with the understanding that a practical implementation of an RME algorithm for current generation Intel multiprocessors is obtained easily by adding a persistence instruction (e.g., calling the pmem_persist function in Intel's Persistent Memory Development Kit (PMDK) [4]) after each memory operation.[3] Instead, we focus our attention on the more fundamental challenges of using memory operations, particularly read-modify-write instructions, in a failure-prone environment: (i) the difficulty of determining during recovery the last memory operation that was performed by a process; and (ii) recovering the response of the memory operation (if applicable), which is returned to the process using a volatile CPU register. These challenges define the character of the RME problem, and establish a common narrative across the system-wide and individual process failure models.

References

1. Andy Rudoff. Programming models for emerging non-volatile memory technologies. *login Usenix Magazine*, 38(3), 2013.
2. Andy Rudoff. Persistent memory programming. *login Usenix Magazine*, 42(2), 2017.
3. Wojciech Golab and Danny Hendler. Recoverable mutual exclusion under system-wide failures. In *Proc. of the 37th ACM Symposium on Principles of Distributed Computing (PODC)*, pages 17–26, 2018.
4. Andy Rudoff and the Intel PMDK Team. Persistent memory development kit, 2020. [last accessed 2/11/2021].

---

[3] The correct use of an RME lock by an application further requires a mechanism to resurrect failed processes. We view the task of implementing such a mechanism as separate from the implementation of the lock.

Prior Work

3

This chapter surveys research results on topics related to recoverable mutual exclusion prior to the publication of Golab and Ramaraju's PODC'16 conference paper [1]. Literature on the more general and widely-studied mutual exclusion problem begins with Dijkstra's seminal paper [2], although the first known solution to the problem is credited to Dekker, who proposed a two-process algorithm that uses one-bit read/write registers. Lamport advanced the state of the art by formalizing the correctness properties of mutual exclusion [3, 4], and also introduced the famous Bakery algorithm as an example of first-come-first-served (FCFS) fairness [5]. Whereas the Bakery uses only reads and writes, and orders processes using numerical tickets, more scalable FCFS algorithms implement various queue structures using read-modify-write primitives [6–9]. For a detailed treatment of progress in mutual exclusion research up to 2003, the reader is referred to [10, 11].

Local spin mutual exclusion algorithms, which guarantee bounded RMR complexity per passage by busy-waiting only on locally accessible shared variables, have been studied intensively due to their performance benefits [6, 12]. For the class of algorithms that use reads, writes and comparison primitives, the tight bound on RMRs per passage in the worst case is $\Theta(\log n)$. Yang and Anderson [13] proved the upper bound in the CC and DSM models using an arbitration tree modeled after Kessels' algorithm [14]. Attiya, Hendler, and Woelfel [15] later proved the matching lower bound, building on a series of earlier results [16–19]. In comparison, queue-based locks achieve $O(1)$ RMR complexity but require additional primitives, such as atomic Fetch-And-Store or Fetch-And-Add.

Recovery from failures is featured prominently in the seminal work of Lamport [4], who formalized two types of faulty process behavior: "unannounced death", similar to a crash in our model but permanent, and "malfunctioning", whereby the private state and communication variables of a process assume arbitrary values. Lamport's Bakery algorithm [5] tolerates the first type of failure provided that a faulty process returns to the non-critical section and its communication variables (i.e., the single-writer shared registers written by

S. Dhoked et al., *Recoverable Mutual Exclusion*, Synthesis Lectures on Distributed Computing Theory, https://doi.org/10.1007/978-3-031-20002-1_3

it) are reset eventually to zero—an assumption that Golab and Ramaraju's model [1, 20] replaces with the requirement that a failed process is resurrected.

Taubenfeld's treatment of fault-tolerant mutual exclusion focuses on a crash-recovery model where process failures affect the values of shared variables in well-defined ways [21]. This model is defined around single-writer registers, and assumes that only those variables that a process "owns" (i.e., has write access to) may be affected by its failure. In one variation, the program counter is reset to the non-critical section on failure, and any variables a process owns are reset to default values. In another variation, the program counter and variables owned by a process may adopt arbitrary values, possibly leading to temporary violations of safety and liveness. Solutions in the latter category are based upon Dijkstra's self-stabilization paradigm [22]. In comparison, a crash failure in a recoverable mutual exclusion algorithm resets the program counter to the non-critical section, and does not affect shared variables as long as they are allocated in PMem.

Bohannon, Lieuwen, Silberschatz, Sudarshan, and Gava [23] proposed a technique for determining the ownership of a Test-And-Set lock following a permanent crash failure, which is the most difficult aspect of recovery in this case given that the mutex algorithm itself is quite simple. In a follow-up paper, Bohannon, Lieuwen, and Silberschatz [24] added recoverability to Mellor-Crummey and Scott's queue-based mutex algorithm [9] by designing an intricate mechanism to repair the queue structure after a process fails while enqueuing or dequeuing itself. Both papers augment the mutex lock with additional shared variables to detect the intent of a process to enter the CS, and perform corrective actions inside a dedicated recovery process that is able to detect failures by querying the operating system. This recovery process is itself assumed to be reliable. Michael and Kim proposed another type of fault-tolerant mutex lock in which a process that is waiting to acquire a lock can "usurp" the lock if it determines that the previous lock holder has crashed permanently [25]. Similarly to [23, 24], this work assumes the ability to detect a crash failure.

In terms of RMR complexity, the recoverable MCS lock presented in [24] does not bound the number of RMRs per passage in the CC or DSM models. This is because when the recovery process is executing corrective actions, a process in the entry section or exit section waits at specific points for such actions to finish by spinning on a global variable. The RMR complexity of this busy-wait loop is unbounded in the DSM model because all processes share the same spin variable. The RMR complexity is also unbounded in the CC model unless the number of failures is bounded, as otherwise the recovery section may invoke corrective actions arbitrarily many times in parallel with one execution of the entry or exit section, each time causing an RMR.

Research on mutual exclusion, and more generally on concurrent objects, has focused mostly on models where the memory is reliable. In this body of work, a limited form of resilience against unreliable processes follows immediately from liveness guarantees in an asynchronous environment, where a slow process cannot be distinguished from one that has crashed permanently. For example, Herlihy introduced *wait-free* objects [26], which guarantee the progress of each correct process individually. In comparison, only a handful of papers consider computation using unreliable memory, focusing on minor corruptions such as bit flips. Afek, Greenberg, Merritt, and Taubenfeld [27] considered the consensus problem in this general context, Moscibroda and Oshman [28] focused on mutual exclusion, and Jayanti, Chandra, and Toueg [29] proposed implementations of shared objects from unreliable base objects. In contrast to these techniques, which break if the number of corruptions exceeds a specified bound, Hoepman, Papatriantafilou, and Tsigas [30], as well as Johnen and Higham [31], proposed self-stabilizing shared objects that can tolerate any number of memory failures but may lose their safety properties temporarily after a failure.

The recoverable mutual exclusion (RME) problem, as formalized by Golab and Ramaraju [1], is a theoretical take on earlier practical work on crash-tolerant locks [23–25]. Its distinguishing feature is the modelling assumption that recovery is empowered by resurrecting a failed process, rather than by relying on system support for precise failure detection, or on a dedicated recovery process. The first attempt to formalize and solve the RME problem appears in the Master's thesis of Aditya Ramaraju [32]. Ramaraju's algorithm, called the RGLock, is a recoverable adaptation of Mellor-Crummey and Scott's queue lock [9] that incurs $O(n)$ RMRs per passage in the worst case for n processes, and $O(1)$ per passage in the absence of failures. The main technical idea in this algorithm is to replace the FAS instruction in the entry section of the MCS lock with a more robust FASAS (Fetch-And-Store-And-Store) instruction that saves the fetched value in shared memory, ensuring that an enqueued process can reliably determine its predecessor in the queue upon recovery from a crash. The FASAS instruction is not supported by modern hardware, and was introduced in [32] specifically to simplify the recovery logic. Recoverable queue locks based on commonly supported primitives are discussed in Chap. 6, and are substantially more complex.

References

1. Wojciech Golab and Aditya Ramaraju. Recoverable mutual exclusion. In *Proc. of the 35th ACM Symposium on Principles of Distributed Computing (PODC)*, pages 65–74, 2016.
2. Edsger W. Dijkstra. Solution of a problem in concurrent programming control. *Communications of the ACM (CACM)*, 8(9):569, 1965.
3. Leslie Lamport. The mutual exclusion problem: part I – a theory of interprocess communication. *Journal of the ACM (JACM)*, 33(2):313–326, 1986.
4. Leslie Lamport. The mutual exclusion problem: part II – statement and solutions. *Journal of the ACM (JACM)*, 33(2):327–348, 1986.
5. Leslie Lamport. A new solution of Dijkstra's concurrent programming problem. *Communications of the ACM (CACM)*, 17(8):453–455, 1974.
6. Thomas E. Anderson. The performance of spin lock alternatives for shared-memory multiprocessors. *IEEE Transactions on Parallel and Distributed Systems (TPDS)*, 1(1):6–16, 1990.
7. Peter Magnusson, Anders Landin, and Erik Hagersten. Queue locks on cache coherent multiprocessors. In *Proc. of the 8th International Parallel Processing Symposium (IPPS)*, pages 165–171, 1994.

8. Gary Graunke and Shreekant Thakkar. Synchronization algorithms for shared-memory multi-processors. *IEEE Computer*, 23(6):60–69, 1990.

9. John M. Mellor-Crummey and Michael L. Scott. Algorithms for scalable synchronization on shared-memory multiprocessors. *ACM Transactions on Computer Systems (TOCS)*, 9(1):21–65, 1991.

10. James H. Anderson, Yong-Jik Kim, and Ted Herman. Shared-memory mutual exclusion: major research trends since 1986. *Distributed Computing (DC)*, 16(2-3):75–110, 2003.

11. Michel Raynal. *Algorithms for Mutual Exclusion*. MIT Press, 1986.

12. Tudor David, Rachid Guerraoui, and Vasileios Trigonakis. Everything you always wanted to know about synchronization but were afraid to ask. In *Proc. of the 24th ACM SIGOPS Symposium on Operating Systems Principles (SOSP)*, pages 33–48, 2013.

13. Jae-Heog Yang and James H. Anderson. A fast, scalable mutual exclusion algorithm. *Distributed Computing (DC)*, 9(1):51–60, 1995.

14. Joep L. W. Kessels. Arbitration without common modifiable variables. *Acta Informatica*, 17:135–141, 1982.

15. Hagit Attiya, Danny Hendler, and Philipp Woelfel. Tight RMR lower bounds for mutual exclusion and other problems. In *Proc. of the 40th ACM Symposium on Theory of Computing (STOC)*, pages 217–226, 2008.

16. James H. Anderson and Yong-Jik Kim. An improved lower bound for the time complexity of mutual exclusion. *Distributed Computing (DC)*, 15(4):221–253, 2002.

17. Robert Cypher. The communication requirements of mutual exclusion. In *Proc. of the 7th ACM Symposium on Parallel Algorithms and Architectures (SPAA)*, pages 147–156, 1995.

18. Rui Fan and Nancy Lynch. An $\Omega(n \log n)$ lower bound on the cost of mutual exclusion. In *Proc. of the 25th ACM Symposium on Principles of Distributed Computing (PODC)*, pages 275–284, 2006.

19. Wojciech Golab, Vassos Hadzilacos, Danny Hendler, and Philipp Woelfel. RMR-efficient implementations of comparison primitives using read and write operations. *Distributed Computing (DC)*, 25(2):109–162, 2012.

20. Wojciech Golab and Aditya Ramaraju. Recoverable mutual exclusion. *Distributed Computing (DC)*, 32(6):535–564, 2019.

21. Gadi Taubenfeld. *Synchronization Algorithms and Concurrent Programming*. Prentice Hall, 2006.

22. Edsger W. Dijkstra. Self-stabilizing systems in spite of distributed control. *Communications of the ACM (CACM)*, 17(11):643–644, 1974.

23. Philip Bohannon, Daniel Lieuwen, Avi Silberschatz, S. Sudarshan, and Jacques Gava. Recoverable user-level mutual exclusion. In *Proc. of the 7th IEEE Symposium on Parallel and Distributed Processing (SPDP)*, pages 293–301, 1995.

24. Philip Bohannon, Daniel Lieuwen, and Avi Silberschatz. Recovering scalable spin locks. In *Proc. of the 8th IEEE Symposium on Parallel and Distributed Processing (SPDP)*, pages 314–322, 1996.

25. Maged M. Michael and Yong-Jik Kim. Fault tolerant mutual exclusion locks for shared memory systems, 2009. US Patent 7,493,618.

26. Maurice Herlihy. Wait-free synchronization. *ACM Transactions on Programming Languages and Systems*, 13(1):124–149, 1991.

27. Yehuda Afek, David S. Greenberg, Michael Merritt, and Gadi Taubenfeld. Computing with faulty shared objects. *Journal of the ACM (JACM)*, 42(6):1231–1274, 1995.

28. Thomas Moscibroda and Rotem Oshman. Resilience of mutual exclusion algorithms to transient memory faults. In *Proc. of the 30th ACM Symposium on Principles of Distributed Computing (PODC)*, pages 69–78, 2011.

29. Prasad Jayanti, Tushar Deepak Chandra, and Sam Toueg. Fault-tolerant wait-free shared objects. *Journal of the ACM (JACM)*, 45(3):451–500, 1998.

30. Jaep-Henk Hoepman, Marina Papatriantafilou, and Philippas Tsigas. Self-stabilization of wait-free shared memory objects. In *Proc. of the 9th International Workshop on Distributed Algorithms (WDAG)*, pages 273–287, 1995.

31. Colette Johnen and Lisa Higham. Fault-tolerant implementations of regular registers by safe registers with applications to networks. In *Proc. of 10th International Conference of Distributed Computing and Networking (ICDCN)*, pages 337–348, 2009.

32. Aditya Ramaraju. RGLock: Recoverable mutual exclusion for non-volatile main memory systems. Master's thesis, University of Waterloo, 2015.

Problem Formulation

4

Recoverable Mutual Exclusion (RME) is a generalization of Dijkstra's mutual exclusion (ME) problem [1] that accommodates certain types of process failures. The original formulation of the RME problem by Golab and Ramaraju [2] considers a model with a fixed set of n asynchronous processes that communicate by accessing variables in shared memory. The processes are labelled, typically using consecutive numerical identifiers as $p_1, p_2, \ldots, p_n$, and each process knows both its own identifier and the constant n. These processes compete for an exclusive lock (mutex), which can be used to protect a shared resource, by following the execution path illustrated in Fig. 4.1. At initialization, as well as immediately after crashing, a process is in the non-critical section (NCS), where it does not access the lock. Upon leaving the NCS, a process always executes the *recovery section*, denoted by the procedure Recover. This code is responsible for repairing the internal structure of the lock, which is potentially corrupted due to earlier crash failures. Next, a process attempts to acquire the lock in the *entry section*, denoted by the procedure Enter. The entry section is sometimes modeled in two parts: a bounded[1] section of code called the *doorway* that determines the order in which processes acquire the lock, followed by a *waiting room*, where processes wait for their turn. Upon completing the entry section, a process has exclusive accesses to the *critical section* (CS). A process subsequently releases the lock by executing the *exit section*, denoted by the procedure Exit, and finally transitions back to the NCS.

The observant reader may question why Recover and Enter are defined as separate procedures given that each process always executes both sequentially. The distinction often follows naturally from the structure of the RME algorithm, where the sections of code responsible for recovery and entry into the CS are clearly delineated. In some algorithms that enforce entry into the CS in a fair order, the boundary between the recovery and entry

---

[1] The term *bounded* in reference to a piece of code means that there exists a function f of the number of processes n such that the code performs at most $f(n)$ shared memory operations in all executions of the algorithm instantiated for n processes.

© The Author(s), under exclusive license to Springer Nature Switzerland AG 2023
S. Dhoked et al., *Recoverable Mutual Exclusion*, Synthesis Lectures on
Distributed Computing Theory, https://doi.org/10.1007/978-3-031-20002-1_4

Fig. 4.1 Execution path of a process participating in recoverable mutual exclusion. A crash failure reverts a process back to the NCS

loop forever
> Non-Critical Section (NCS)
> `Recover()`
> `Enter()` $\left\{\begin{array}{l}\text{Doorway}\\\text{Waiting Room}\end{array}\right.$
> Critical Section (CS)
> `Exit()`

Fig. 4.2 Alternative execution path of recoverable mutual exclusion with a combined recovery/entry section

loop forever
> Non-Critical Section (NCS)
> `Recover-Enter()`
> Critical Section (CS)
> `Exit()`

sections is determined more precisely by the position of the doorway, as shown in Fig. 4.1. That is, the recovery section ends prior to the first step of the doorway, which defines the start of the entry section, as stipulated by Golab and Ramaraju in [3]. Thus, the doorway remains a prefix of the entry section, as in conventional mutual exclusion, even in algorithms that perform unbounded recovery actions prior to executing the doorway. On the other hand, the recovery section has no formal significance in algorithms that lack both a doorway and specialized recovery code. In practice, programmers may choose to implement `Recover` and `Enter` as one combined procedure for convenience, allowing the application to acquire the recoverable mutex lock using only a single procedure call. We will denote such a procedure as `Recover-Enter`, as shown in Fig. 4.2.

The exact position of the doorway in an algorithm is a matter of independent interest. In conventional ME, the doorway is defined as a bounded prefix of the entry section [4], and so its position is effectively determined by its last instruction. The latter is uniquely defined if we assume that the doorway is a *minimal* prefix, meaning that it comprises the smallest possible number of instructions. In comparison, the situation is less constrained in the RME problem because the start of the doorway is no longer anchored to the first instruction a process executes upon leaving the non-critical section. Moreover, an RME algorithm may execute unbounded synchronization code (e.g., a busy-wait loop) both before and after the doorway. That said, inappropriate definitions of the doorway that would make the recovery section or entry section trivial are ruled out in two ways. First, recovery code must be bounded in the absence of failures (see Sect. 4.3), and so the doorway cannot begin arbitrarily late in the algorithm. This rules out the case where unbounded synchronization is pushed into the recovery section to such extent that the entry section is reduced to a single instruction—a trivial doorway followed by an empty waiting room. Second, if we assume once again that the doorway is minimal, then any recovery code that does not affect the order of entry into the CS is excluded from the entry section, and hence the doorway cannot begin arbitrarily early in the algorithm. This rules out the case where bounded recovery code is pushed into the entry section to such extent that the recovery section becomes empty.

4.1 Modeling Steps and Execution Histories

Correctness properties for mutex algorithms are expressed in reference to *histories* that record the actions of processes as they execute the pseudo-code of Fig. 4.1. Formally, a history H is a sequence of *steps* that come in two varieties: *ordinary steps* and *crash steps*. An ordinary step is a shared memory operation combined with local computation such as arithmetic operations, accessing one or more private variables, and advancing the program counter. A crash step denotes either the failure of an individual process, or the simultaneous failure or all processes. For each affected process, the crash step resets the private variables of that process to their initial values, while leaving the values of shared variables intact.[2] These private variables include a per-process *program counter*, which identifies the next ordinary step a process is poised to take. The program counter points to the NCS initially, and is updated in each step. We say that a process is *at line X* of an algorithm if its program counter identifies an ordinary step corresponding to line X of the algorithm's pseudo-code. One line of pseudo-code may entail multiple steps.

As hinted earlier, a crash step resets the program counter of the failed process back to the NCS. The next step a process takes after a crash is either another crash step, or the first step of Recover.[3] A process is said to *recover* following a crash step by executing the first step of Recover. A *passage* is a sequence of steps taken by a process from when it begins Recover to when it completes Exit, or crashes, whichever event occurs first. A passage is called *failure-free* if it does not end with a crash step, which includes all incomplete passages. A *super-passage* is a maximal non-empty collection of consecutive passages executed by the same process where (only) the last passage in the collection is failure-free. A process is *executing a super-passage* if it is either outside the NCS, or in the NCS following a crash failure. A process may execute the CS at most once per passage, and possibly multiple times in one super-passage.

A process that completes the CS and then crashes in Exit is required in the Golab and Ramaraju model [2, 3] to continue taking steps until it completes a failure-free passage, which

---

[2] In practical terms, it is assumed that shared variables are stored in main memory and read back from this memory during recovery from a crash, in contrast to private variables, which are always reinitialized.

[3] The statements in this paragraph assume that the history records steps for only a single RME lock. In scenarios where locks are nested, which we describe later on in Sect. 4.2, the analogous statements apply to the projection of the history onto any one of the locks.

entails one or more additional and redundant executions of the CS. Although this behavior is somewhat inefficient, it simplifies the model conceptually by reducing the number of possible execution paths. Alternative models that allow more flexible state transitions are discussed in Sect. 4.2.

The set of histories generated by possible executions of a mutex algorithm is prefix-closed. For any finite history H, we denote the length of H (i.e., number of steps in H) by $|H|$. A history H is *fair* if it is finite, or if it is infinite and every process either executes zero steps, or halts in the NCS after completing a failure-free passage, or executes infinitely many steps. We assume for simplicity of analysis that the critical section is bounded. Intuitively, we expect RME algorithms with bounded critical sections to guarantee certain liveness properties in fair histories. As explained shortly in Sect. 4.3, liveness properties must be conditioned in some way on the sparsity of crash steps in a history, in addition to fairness, as otherwise recurring failures may cause processes to transition perpetually to the non-critical section without ever reaching the critical section.

Aside from hindering entry into the critical section, crash failures may have indirect effects on the execution paths of processes. For example, a process that is executing a new passage after recovering from a crash may need to perform additional actions to repair the internal state of the recoverable lock, even if this new passage is itself failure-free. These actions, in turn, may affect the execution paths of other processes, including ones that have never failed! Thus, processes may incur penalties (e.g., in terms of time complexity) in a given passage due to an earlier failure of the same process, or even the failure of another process. Modelling such interactions precisely requires additional terminology beyond simply classifying passages as failure-free or not. As a first step toward this goal, Golab and Ramaraju [2] define a process *in cleanup* as one executing `Recover` following a passage that ended with a crash step, and make reference to passages that *begin in cleanup*. Intuitively, such passages are most directly influenced by process crashes. To capture more subtle effects, Golab and Ramaraju [3] formalize the notion that one passage may exert indirect influence on another using the concept of interference among passages, and k-failure-concurrency:

Definition 4.1 A passage *interferes* with another passage (of the same process or a different one) if their corresponding super-passages are concurrent, meaning that neither super-passage ends (in a failure-free passage) before the other begins.

Different degrees of influence among passages are then formalized by iterating the above binary relation.

Definition 4.2 For any integer $k \geq 0$, a passage is *k-failure-concurrent* (or k-FC) if and only if:

- $k = 0$ and the passage begins in cleanup or ends with a crash (i.e., the process executing the passage crashes in its corresponding super-passage); or
- $k > 0$ and the passage interferes with any passage (including possibly itself) that is $(k - 1)$-failure-concurrent.

Figure 4.3 illustrates how Definition 4.2 is applied. The two passages of process p belong to the same super-passage, and both are 0-FC because p crashes in its first passage and then begins in cleanup in its second passage. Process q does not crash, and so each of its super-passages is one passage long. The first two passages of q are 1-FC because they interfere with p's 0-FC passages, but q's third passage is not k-FC for any k because it begins after p's (one and only) super-passage ends. Intuitively, the 0-FC passages of p are affected most directly by p's crash, whereas q's 1-FC passages may be affected indirectly. For example, p's crash may cause an RME algorithm to open a special recovery path, which may inflate the RMR complexity for q temporarily.

It follows easily that if a passage is k-FC then it is also k'-FC for all $k' > k$. As a result, if an algorithm satisfies some property in all k-FC passages for some specific $k \geq 0$, then it also satisfies the same property in any k''-FC passage for $0 \leq k'' < k$, because any such k''-FC passage is also k-FC. In general, when a desirable property holds *except* in k-FC passages, it is preferred for the parameter k to be as small as possible since that minimizes the subset of passages that are exempt from the desirable behaviour. The same comment applies when an undesirable property holds *only* in k-FC passages. This principle is illustrated in Fig. 4.4, where the desirable property is $O(1)$ RMR complexity and the undesirable property is $O(n)$ RMR complexity. Figure 4.4a shows a hypothetical algorithm where the desirable property holds except in 0-FC passages (smaller k, superior algorithm), and Fig. 4.4b shows an algorithm where the same property holds except in 1-FC passages (larger k, inferior algorithm).

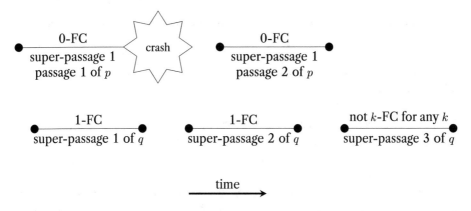

Fig. 4.3 Examples of failure concurrency among passages of processes p and q

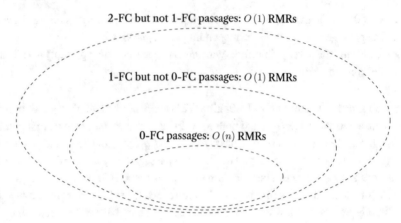

a) Superior algorithm guarantees $O(1)$ RMR complexity except in 0-FC passages.

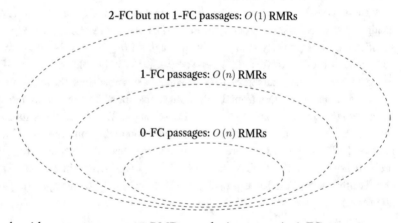

b) Inferior algorithm guarantees $O(1)$ RMR complexity except in 1-FC passages.

Fig. 4.4 Venn diagrams showing RMR complexity of passages at different degrees of failure concurrency

4.2   Control Flow

The model of Golab and Ramaraju [2, 3] intentionally defines a linear flow of control in which processes always execute code in the order prescribed by Fig. 4.1—NCS, `Recover`, `Enter`, CS, and `Exit`–as opposed to allowing branches from `Recover` to the section where a crash actually occurred. This convention follows naturally from the notion that `Recover`, `Enter` and `Exit` are procedures that are called by an application, and are punctuated by application-specific code (e.g., the NCS and CS), rather than flat sections of code embedded in a contiguous program. Branches from the code of one procedure to the code of a different procedure are not permitted because, in practice, one procedure must return control to the application (i.e., pop a stack frame) before the application invokes the next procedure in the sequence. For example, a process that has no recovery actions to perform can branch from the beginning of Recover to the end of the same procedure, skipping the body of the recovery section, but a process recovering from a failure that occurred in Exit cannot branch from Recover (i.e., from code internal to the recovery section) directly to Exit and skip over the application-specific CS code. In fact, such a branch would cause a compilation error if implemented using a goto statement in C or C++. Alternative formulations of the RME problem that allow more flexible transitions between different sections of code are discussed later on in Sect. 4.5.

Some implementations of RME locks allow processes to deviate from the linear flow of control in one important case, namely when a process crashes inside the CS. In this special situation, the faulty process p may be able to resume execution directly inside the CS on recovery, rather than transitioning to the NCS first and re-entering the CS in the usual manner by completing the recovery and entry sections. In practical terms, this means that p's algorithm skips Recover and Enter on recovery, and transfers control directly to some point in the CS (e.g., to the line of code immediately following the call to Enter). We emphasize that the decision to bypass the recovery and entry sections is made by the application program that uses an RME lock, and *not* by code inside the lock's Recover and Enter procedures. We will refer to the RME lock's ability to tolerate such application-initiated shortcuts as *CS continuity*.[4] Informally speaking, this property is supported automatically by any RME lock where p's crash cannot be detected, for example by querying a failure detector or by reading program variables, by p itself or by another process inside the Recover, Enter, and Exit procedures. With the exception of Chap. 11, which considers system-wide process failures,[5] all of the RME algorithms discussed in this book allow CS continuity as they do not use failure detectors or special variables whose state could reveal a crash.

Figure 4.5 is a simple example of how a process could access an RME lock, and also recover from a failure. The process begins execution at line 1, which is part of the NCS with respect to *lock A*, then acquires and releases the lock at lines 2 to 5. A crash failure restarts the program from the beginning at line 1, and thus resets the program counter to a line of code inside the non-critical section of *lock A*. The initialization code at line 1 decides how to recover the execution, for example by repeating lines 2 to 5 , or by executing a specialized recovery procedure that recovers *lock A* by calling *lock A*.Recover, *lock A*.Enter, and

---

[4] CS continuity is orthogonal to *Critical Section Re-entry (CSR)* [2], which we formalize shortly. CSR entails re-entering the CS in the usual manner, by returning to the NCS first (hence releasing the CS), and executing the recovery and entry sections.

[5] CS continuity is fundamentally incompatible with the system-wide failure model because when a process p fails in the CS, it possible that another process q fails simultaneously in Recover, Enter, or Exit, in which case q may be able to detect the failure by reading program variables.

Fig. 4.5 Example of control
flow, single lock

```
1 ... //  lockA non-critical section
2 lockA.Recover()
3 lockA.Enter()
4 ... //  lockA critical section
5 lockA.Exit()
6 ... //  lockA non-critical section
```

$lockA$.Exit.[6] In either case, the procedures of $lockA$ must be accessed according to the specification of the RME problem, meaning that the first access to $lockA$ after the crash is a call to Recover.

As hinted earlier, some RME lock implementations support CS continuity and allow a process that crashes inside the critical section of $lockA$ at line 4 to bypass Recover and Enter during recovery. For example, this could be accomplished at line 1 of Fig. 4.5 by analyzing the values of program variables to detect the crash failure, and then branching from line 1 directly to line 4.

Figures 4.6 and 4.7 illustrate recovery in more complex programs where multiple RME locks ($lockA$ and $lockB$) are held concurrently, and where improper recovery actions can potentially cause deadlock. We will assume in both examples that lock ordering is used for deadlock prevention, meaning informally that $lockA$ is always acquired before $lockB$. As we discuss shortly, this requirement sometimes additionally implies that any recovery actions with respect to $lockB$ must be completed before any process attempts to recover $lockA$, meaning that the two locks are recovered in reverse of their acquisition order. Failure to observe this principle can lead to a situation where one process p recovers $lockA$ and proceeds to recover $lockB$ while another process q that held $lockB$ at the time of failure is blocked on $lockA$.Recover, in which case p may wait forever in $lockB$.Recover because q never reaches $lockB$.Recover and thus deviates from the execution flow defined in Fig. 4.1 with respect to $lockB$ (i.e., q crashes and remains in the NCS of $lockB$ forever instead of eventually executing $lockB$.Recover).

In Fig. 4.6, $lockB$ is nested inside $lockA$, meaning that $lockA$ is acquired before $lockB$ and $lockB$ is released before $lockA$. Such a composition could occur naturally in a tournament tree where $lockA$ is a leaf-level lock and $lockB$ is its parent. As in Fig. 4.5, a crash failure restarts the program from the beginning at line 7, which is the non-critical section with respect to both $lockA$ and $lockB$. In the simplest case, the entire program is repeated following recovery. For example, to recover from a crash at line 16, which is in the exit section of $lockA$ and non-critical section of $lockB$, a process would acquire $lockA$ at lines 8 to 9 , acquire $lockB$ at lines 11 to 12, release $lockB$ at line 14, and finally release $lockA$ at line 16. Alternatively, a specialized recovery procedure could be executed that acquires

---

[6] Such a passage through $lockA$ serves two goals: to recover the internal structure of $lockA$, and to complete or roll back any interrupted actions corresponding to line 4 in Fig. 4.5. The second goal is optional, and if only the first goal is relevant then $lockA$.Exit and $lockA$.Enter are executed back-to-back with an empty CS.

```
 7 ... // lockA non-critical section, lockB non-critical section
 8 lockA.Recover()
 9 lockA.Enter()
10 ... // lockA critical section, lockB non-critical section
11 lockB.Recover()
12 lockB.Enter()
13 ... // lockA critical section, lockB critical section
14 lockB.Exit()
15 ... // lockA critical section, lockB non-critical section
16 lockA.Exit()
17 ... // lockA non-critical section, lockB non-critical section
```

Fig. 4.6 Example of control flow, nested locks

```
18 ... // lockA non-critical section, lockB non-critical section
19 lockA.Recover()
20 lockA.Enter()
21 ... // lockA critical section, lockB non-critical section
22 lockB.Recover()
23 lockB.Enter()
24 ... // lockA critical section, lockB critical section
25 lockA.Exit()
26 ... // lockA non-critical section, lockB critical section
27 lockB.Exit()
28 ... // lockA non-critical section, lockB non-critical section
```

Fig. 4.7 Example of control flow, hand-over-hand locking

and releases $lockA$ only, since the failure occurred in the non-critical section of $lockB$. Deadlock on recovery, via the mechanism explained earlier, becomes a concern only if a process crashes during a passage through $lockB$ at lines 11 to 14, which are inside the CS of $lockA$. In summary, the recovery of $lockA$ by some process p may occur in a state where $lockB$ is effectively held by some process q that has transitioned back to the NCS of both locks, and where p cannot progress further until q reacquires $lockA$. To avoid deadlock, either $lockB$ should be recovered first, or q should be permitted to re-enter the CS of $lockA$ in a bounded number of its own steps (ahead of p) so that it can recover $lockB$ instead of waiting on $lockA$ behind p. We formalize this ability to re-enter the CS later on in this chapter as *bounded critical section re-entry*.

Additional recovery paths in Fig. 4.6 are possible if the RME locks support CS continuity. For example, to recover from a crash at line 12, which is in the critical section of $lockA$ and entry section of $lockB$, a process could branch from line 7 to line 10 or line 11, which effectively resumes execution in the critical section of $lockA$ and non-critical section of $lockB$, and continue executing lines 11 to 16. The same options apply following a crash at line 14, which is in the critical section of $lockA$ and exit section of $lockB$. Similarly, if a

Fig. 4.8 Example recovery
procedure for the program in
Fig. 4.7

29 *lock B*.Recover()
30 *lock B*.Enter()
    // empty critical section
31 *lock B*.Exit()

process crashes at line 13, which is in the critical section of both *lock A* and *lock B*, it could
recover by branching from line 7 back to line 13.

Figure 4.7 illustrates an alternative form of lock composition where *lock A* is released after
lock B is acquired, but before *lock B* is released. This pattern, known to practitioners as *hand-
over-hand locking* or *lock coupling*, is used sometimes to traverse linked data structures.
Compared to Fig. 4.6, which can be recovered correctly by restarting from the beginning
as long as *lock A* allows a process to re-enter the CS after a failure, Fig. 4.7 is more prone
to deadlock as a process may crash while holding *lock B* but not *lock A*, such as at lines 25
to 27. In this situation, *lock B* must be recovered and released explicitly before *lock A* is
accessed again during recovery, because ownership of *lock A* may have already been passed
to another process that is waiting to acquire *lock B*. Therefore, the recovery procedure should
first make a complete passage through *lock B*, as shown in Fig. 4.8, except possibly in the
special case when the application is able to rule out a failure at lines 25 to 27 . After that,
the entire program can be repeated from line 18, and shortcuts may be applicable if the
crash occurred at specific places (e.g., line 21, line 24, or line 26) and the locks support CS
continuity.

In situations where more than two locks are accessed in a hand-over-hand manner, for
example when traversing a long linked list, the recovery technique described above remains
applicable as long as the program remembers which two RME locks it acquired most recently
prior to the failure. Supposing that these locks were *lock A* and *lock B*, in order of acquisition,
the recovery procedure could complete one passage through *lock B* to recover and release it,
then do the same to *lock A*, and then restart the entire traversal from the beginning of the list.
In practical terms, this requires storing persistent pointers to the last two linked list nodes
traversed, and ensuring that these nodes are not deleted and reclaimed before the recovery
procedure runs to completion.

4.3   Correctness Properties

The correctness properties of recoverable mutual exclusion are derived from Lamport's
formalism [4], which considers only permanent crashes ("unannounced death") and Byzan-
tine failures ("malfunctioning"). Since multiple alternative definitions of the correctness
properties for RME have appeared in the literature, we begin this section with the original
problem specification of Golab and Ramaraju [2, 3]. Fundamental differences between the
correctness properties for RME and Lamport's definitions are highlighted below in *italics*.
The word *terminates*, when used in reference to an execution of Recover, Enter, CS or

Exit by some process, refers to the corresponding section either being executed to completion without failure, or being executed partially and then interrupted by a crash failure of the executing process.

Definition 4.3 *(Mutual Exclusion (ME))* For any finite history H, at most one process is in the CS at the end of H.

Definition 4.4 *(Deadlock-Freedom (DF))* For any infinite fair history H *with finitely many failures*, if a process p_i leaves the NCS in some step of H then eventually some process p_j (possibly $j \neq i$) is in the CS.

Definition 4.5 *(Starvation-Freedom (SF))* For any infinite fair history H *with finitely many failures*, if a process p_i leaves the NCS in some step of H then eventually p_i itself enters the CS.

Definition 4.6 *(Terminating Exit (TE))* For any infinite fair history H *with finitely many failures*, any execution of Exit by a process p_i terminates in a finite number of p_i's steps.

Definition 4.7 *(Bounded Exit (BE))* For any history H, any execution of Exit by a process p_i terminates in a bounded number of p_i's steps.

As hinted earlier in Sect. 4.1, liveness properties are conditioned on sparsity of failures, and this is indeed the main difference between the specifications of RME and classic ME. Specifically, the DF, SF, and TE properties for RME guarantee progress explicitly only in histories containing finitely many failures. It is important to note, however, that the same properties also imply progress in fair histories that contain infinitely many failures, provided that such failures are sufficiently sparse. For example, the definition of starvation freedom implies that p_i eventually enters the CS as long as there are finitely many failures *while p_i is on the path to the CS after leaving the NCS*, even if there are infinitely many additional failures after p_i finally enters the CS. Similarly, the definition implies that p_i eventually enters the CS as long as for any integer $c > 0$, the fair history eventually has a failure-free fragment F where every process that does not halt in the NCS after a failure-free passage takes at least c steps.

Beyond the fundamental safety and liveness properties, several additional definitions are introduced in [2, 3] to constrain the time complexity of the recovery section.

Definition 4.8 *(Bounded Recovery (BR)* For any history H, any execution of Recover by a process p_i terminates in a bounded number of p_i's steps.

Definition 4.9 *(k-Bounded Recovery (k-BR))* For any history H, any execution of Recover by a process p_i terminates in a bounded number of p_i's steps unless p_i's passage is k-failure-concurrent.

It is desirable, and arguably essential, for a recoverable mutual exclusion algorithm to satisfy BR or k-BR for some $k \geq 0$, which ensures that the recovery section is bounded in failure-free executions. It follows easily that BR is strictly stronger than 0-BR, and that for any $k \geq 0$ and $k' > k$, k-BR is strictly stronger than k'-BR. In any algorithm that satisfies BR, the recovery code can be shifted to the entry section and joined with the doorway, although doing so may affect some other correctness properties by lengthening the doorway unnecessarily. For example, first-come-first-served (FCFS) fairness may be weakened as a result.

Two other RME-specific definitions are proposed in [2, 3] to simplify recovery from a crash failure inside the critical section, and make it easier to nest recoverable mutex locks:

Definition 4.10 *(Critical Section Re-Entry (CSR))* For any history H and for any process p_i, if p_i crashes inside the CS then no other process may enter the CS before p_i re-enters the CS after the crash failure under consideration.

Definition 4.11 *(Bounded Critical Section Re-Entry (BCSR))* For any history H and for any process p_i, if p_i crashes inside the CS then p_i incurs a bounded number of steps in each subsequent execution of the recovery and entry sections until it re-enters the CS.

It is straightforward to show that BCSR implies CSR, and this point is proved formally in [3].

The last set of definitions we discuss pertains to first-come-first-served (FCFS) fairness, which takes on a new flavour in the context of RME due to the subtle effects of failures. The standard definition of FCFS is given first for completeness:

Definition 4.12 *(First-Come-First-Served (FCFS))* For any history H, suppose that process p_i completes the doorway in its ℓ_i-th passage before process p_j begins the doorway in its ℓ_j-th passage. Then p_j does not enter the CS in its ℓ_j-th passage before p_i enters the CS in its ℓ_i-th passage.

The above definition may be achievable in some RME algorithms, but is fundamentally incompatible with the CSR property. Consider a situation where p_i completes the doorway in passage ℓ_i before p_j begins the doorway in passage ℓ_j, then p_j completes the doorway in passage ℓ_j, then p_i enters the CS and crashes while p_j is still in the entry section. Upon recovery of p_i, CSR requires that no process enters the CS before p_i does so in passage $\ell_i + 1$, and this includes p_j in passage ℓ_j since we assume that p_j was before the CS in passage ℓ_j at the time when p_i crashed. On the other hand, the classic FCFS property forbids p_i in passage $\ell_i + 1$ from entering the CS before p_j in passage ℓ_j because p_j in passage ℓ_j completed the doorway before p_i began passage $\ell_i + 1$, and hence before p_i began the doorway in passage $\ell_i + 1$.[7] Thus, the requirements of CSR and FCFS are in contradiction in this example. To address this problem, Golab and Ramaraju [3] propose an alternative fairness property called k-FCFS.

Definition 4.13 (*k-First-Come-First-Served (k-FCFS)*) For any history H, suppose that process p_i begins its ℓ_i-th passage and p_j begins its ℓ_j-th passage. Suppose further that neither passage is k-failure-concurrent, and that process p_i completes the doorway in its ℓ_i-th passage before process p_j begins the doorway in its ℓ_j-th passage in H. Then p_j does not enter the CS in its ℓ_j-th passage before p_i enters the CS in its ℓ_i-th passage.

Intuitively, k-FCFS is a weakening of FCFS that ignores k-failure-concurrent passages, where processes may enter the CS out of order with respect to their execution of the doorway. In fact, the doorway can be undefined or unbounded in such passages since it is not relevant. It follows easily that FCFS is strictly stronger than 0-FCFS, and that for any $k \geq 0$ and $k' > k$, k-FCFS is strictly stronger than k'-FCFS.

4.4   Alternative Definitions of Correctness Properties

Several alternative definitions of correctness properties for RME have appeared in the literature since Golab and Ramaraju [2], particularly properties related to fairness. This section covers some of these variations.

4.4.1 Starvation Freedom

In the context of RME, to *starve* means to remain forever in the recovery or entry section, for example inside a busy-wait loop, during a single passage. Golab and Ramaraju's definition of starvation freedom, defined earlier in Sect. 4.3 and denoted henceforth as GR-SF, emphasizes simplicity and allows starvation in certain cases despite the system as a whole making progress. This weakness was first documented by Jayanti and Joshi [5], who pointed out that GR-SF allows one process to starve while another processes passes through the CS infinitely often, as long as at least one process fails infinitely often. We illustrate this problematic behaviour in Fig. 4.9, where process p starves, process q completes failure-free passages infinitely often, and process r completes passages infinitely often while also crashing infinitely often. Although q and r make progress, p is denied entry into the CS forever.

---

[7] It is possible that p_i bypasses the doorway entirely in passage $\ell_i + 1$ due to CSR, and therefore never begins the doorway in passage $\ell_i + 1$. Our point holds regardless.

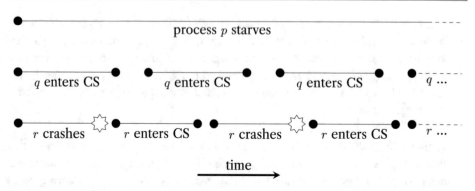

Fig. 4.9 Example of starvation in a hypothetical algorithm that satisfies Golab and Ramaraju's SF property

To remedy the problem with GR-SF, Jayanti and Joshi formulated an alternative definition of starvation freedom that avoids the finite failure assumption, and instead mandates that failures be sparse in a precise sense that is stated below[8]:

Definition 4.14 *(Jayanti and Joshi Starvation-Freedom (JJ-SF))* For any infinite fair history *H where each process fails finitely often in any super-passage*, if a process p_i begins a super-passage then eventually p_i itself enters the CS in that super-passage.

The JJ-SF property eliminates the scenario in Fig. 4.9 since process r crashes only once in any super-passage. Note that such an execution is admissible only in models with independent crash failures, and indeed JJ-SF is equivalent to GR-SF in models with system-wide failures. This holds because GR-SF allows starvation only in executions with infinitely many failures, and no process can possibly starve (in a single passage) when system-wide failures occur infinitely often.

The JJ-SF property rules out one special case of starvation that is not caught by GR-SF, but allows another: a process p_i may starve while another process p_j passes through the CS infinitely often without failing, as long as some other process p_k fails infinitely often in one super-passage. We illustrate this problematic behaviour in Fig. 4.10, where process p starves, process q completes failure-free super-passages infinitely often, and process r crashes infinitely often in a single super-passage. Although q makes progress, p is denied entry into the CS forever, and this behaviour is admissible because of r's failures.

An even stronger notion of fairness that rules out this behaviour is obtained by combining GR-SF with another property formulated by Golab and Hendler [6]:

---

[8] We paraphrase the definition slightly by using the term *super-passage* in place of Jayanti and Joshi's *attempt*. Differences between the Golab-Ramaraju model and Jayanti-Joshi model are discussed further in Sect. 4.5.

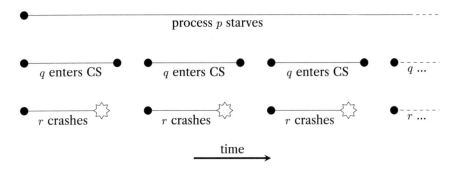

Fig. 4.10 Example of starvation in an algorithm that satisfies Jayanti and Joshi's SF property

Definition 4.15 (*Failures-Robust Fairness (FRF)*) For any infinite fair history H *containing infinitely many super-passages*, if a process p_i leaves the NCS in some step of H then eventually p_i itself enters the CS.

Informally speaking, FRF states that if the system as a whole is making progress in terms of completion of super-passages, then each process individually is guaranteed progress irrespective of the pattern of failures. This eliminates the starvation scenario illustrated in Fig. 4.10, where q completes infinitely many super-passages.

Theorem 4.16 *In a model with independent process failures, the combination of GR-SF and FRF implies JJ-SF.*

Proof Let H be an infinite fair history that satisfies both GR-SF and FRF. Suppose for contradiction that H does not satisfy JJ-SF. Then each process crashes finitely many times per super-passage in H, and yet some process p_i starves. Furthermore, H must contain infinitely many failures, as otherwise GR-SF would rule out the possibility of p_i starving. Thus, some process p_j fails infinitely often in H despite failing finitely many times per super-passage. This implies that p_j completes super-passages infinitely often, and hence H contains infinitely many super-passages. However, in that case the FRF property of H contradicts p_i starving in H. □

4.5   Alternative Problem Definitions

The introduction of distinct recovery and entry sections in the model of Golab and Ramaraju [2, 3] opens the door to alternative definitions of the RME problem that allow non-linear flow of control. In fact, one such variation was already presented by Ramaraju [7] prior to the publication of [2, 3], allowing for transitions directly from Recover to the CS (bypassing Enter), and to Exit (bypassing both Enter and the CS). A similar model was also

Fig. 4.11 Example of control
flow in Katzan and Morrison's
model, single lock

```
32 ... // lockA non-critical section
33 ret := lockA.Recover()
34 if ret = enter then
35  ⌊ lockA.Enter()
36 if ret = enter ∨ ret = CS then
37  ⌊ ... // lockA critical section
38 lockA.Exit()
39 ... // lockA non-critical section
```

adopted later on by Jayanti and Joshi [5]. The intention underlying such non-linear models
is to resume execution from the point of the most recent failure, or else proceed to the entry
section during failure-free execution. In some (but not all) cases, such as a failure in the
exit section, this avoids redundant executions of the entry section and the CS, and reduces
the impact of failures on the algorithm's RMR complexity. Notably, Jayanti and Joshi's
algorithm [5] incurs only $O(1)$ additional RMRs per failure, and in that sense it is superior
to Golab and Ramaraju's algorithms [2, 3] where each failure adds $\Omega(\log n)$ RMRs. On the
other hand, non-linear models in general suffer from an important drawback: they assume
the ability to jump (e.g., via a goto statement) from Recover to the CS, or to specific lines
of code in Enter and Exit, without returning back to Recover. As explained earlier in
this chapter, such jumps are not supported by mainstream procedural and object-oriented
programming languages.

Katzan and Morrison [8] presented a more practical reformulation of the RME problem
where control flow is left up to the application, and is merely guided by the analysis performed
in the recovery section. In their model, Recover returns a three-valued indicator specifying
that control should be transferred to Enter, CS, or Exit, and it is up to the application
to ensure that the correct flow of control is maintained. As an example, Fig. 4.11 illustrates
how this could be accomplished in an application that uses a single lock.[9] Compared to Fig.
4.5, which is the analogous example in the Golab-Ramaraju model, the code in Fig. 4.11
is more complex, but ensures that the application can recover from a crash in $lockA$.Exit
without executing the CS again.

Figures 4.12 and 4.13 illustrate different compositions of locks in the Katzan-Morrison
model, and are analogous to Figs. 4.6 and 4.7. We run into the risk of deadlock again in the
hand-over-hand locking example (Fig. 4.13), and in this case we deal with it in two ways.
First, the recovery sections of both $lockA$ and $lockB$ are executed at lines 56 to 57 before
any attempt is made to reacquire either lock. Second, the body of the algorithm is divided
into two cases: if the process does not hold $lockB$ according to the outcome of line 57 then
it executes lines 59 to 66, where it reacquires $lockA$ (if required) and $lockB$, then releases

---

[9] All of our examples illustrating the Katzan-Morrison model assume for simplicity that the lock is
not abortable.

```
40 ... // lockA non-critical section, lockB non-critical section
41 retA := lockA.Recover()
42 if retA = enter then
43    lockA.Enter()
44 if retA = enter ∨ retA = CS then
45    ... // lockA critical section, lockB non-critical section
46    retB := lockB.Recover()
47    if retB = enter then
48       lockB.Enter()
49    if retB = enter ∨ retB = CS then
50       ... // lockA critical section, lockB critical section
51    lockB.Exit()
52    ... // lockA critical section, lockB non-critical section
53 lockA.Exit()
54 ... // lockA non-critical section, lockB non-critical section
```

Fig. 4.12 Example of control flow in Katzan and Morrison's model, nested locks

lockA; otherwise the process already holds *lockB* and executes only lines 68 to 72, where it releases *lockA* (if required).

4.6   Remote Memory References

Aside from fundamental safety and liveness properties, we are interested in the efficiency of mutex algorithms, particularly with respect to the number of *remote memory references* (RMRs) executed by a process per passage. The precise definition of RMRs depends on the shared memory hardware architecture, and in this work we consider both the cache-coherent (CC) and distributed shared memory (DSM) models [9], which are illustrated in Fig. 4.14. In the CC model, any memory location can be made locally accessible to a process by creating a copy of it in the corresponding processor's cache. A distributed coherence protocol ensures that when a memory location is overwritten by one processor, cached copies held by other processors are either invalidated or updated. In contrast, the DSM model lacks (coherent) caches but allows each processor to locally access a portion of the address space corresponding to its attached memory module.

Loosely speaking, an RMR in the CC or DSM model is any memory operation that traverses the interconnect shown in Fig. 4.14. When calculating upper bounds on RMR complexity in the CC model, we conservatively count each shared memory operation as an RMR with the exception of an *in-cache* read, which occurs when a process p_i reads a variable v that it has already read in an earlier step, following which step no process has accessed

```
55  ... // lockA non-critical section, lockB non-critical section
56  retA := lockA.Recover()
57  retB := lockB.Recover()
58  if retB = enter then
59  |   if retA = enter then
60  |   |   lockA.Enter()
61  |   if retA = enter ∨ retA = CS then
62  |   |   ... // lockA critical section, lockB non-critical section
63  |   lockB.Enter()
64  |   ... // lockA critical section, lockB critical section
65  |   lockA.Exit()
66  |   ... // lockA non-critical section, lockB critical section
67  if retB = CS then
68  |   if retA = CS then
69  |   |   ... // lockA critical section, lockB critical section
70  |   if retA = CS ∨ retA = exit then
71  |   |   lockA.Exit()
72  |   ... // lockA non-critical section, lockB critical section
73  lockB.Exit()
74  ... // lockA non-critical section, lockB non-critical section
```

Fig. 4.13 Example of control flow in Katzan and Morrison's model, hand-over-hand locking

Fig. 4.14 Abstract models of shared memory architectures—DSM (left) and CC (right)

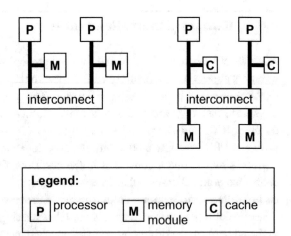

Legend:

P processor   M memory module   C cache

v except by a read operation. For lower bounds, we conservatively count as an RMR any operation that leads to a cache miss, meaning that a process p_i accesses a variable v that is has never accessed before, or that it has accessed earlier and where v was subsequently overwritten by another process. In the DSM model, each shared variable is local to exactly one process (assuming a one-to-one mapping between processes and processors), which is determined statically at initialization. An ME or RME algorithm is called *local spin* if there exists an upper bound on its worst-case RMR complexity per passage. The RME complexity for an RME algorithm can be analyzed more precisely by considering special cases, such as failure-free passages, or passages that are k-failure-concurrent for specific k.

4.7 Synchronization Primitives

Many RME algorithms described in this monograph assume that, in addition to regular read and write instructions, the multiprocessor also supports certain atomic read-modify-write (RMW) instructions. Common examples of such RMR instructions include the following:

A *compare-and-swap* (CAS) instruction takes three arguments: *address*, *old* and *new*; it compares the contents of a memory location (*address*) to a given value (*old*) and, only if they are the same, modifies the contents of that location to a given new value (*new*). It returns true if the contents of the location were modified and false otherwise.

A *fetch-and-store* (FAS) instruction takes two arguments: *address* and *new*; it replaces the contents of a memory location *address* with a given value *new* and returns the old contents of that location.

A *fetch-and-add* (FAA) instruction takes two arguments: *address* and *val*; it adds a given value *val* to the contents of a memory location *address* and returns the old contents of that location.

A *fetch-and-increment* (FAI) instruction takes one argument: *address*; it increments the contents of a memory location *address* by one and returns the old contents of that location.

The above RMW instructions can be defined more formally as the atomic execution of the pseudocode procedures presented in Fig. 4.15. In practice, the return values of these procedures are returned to a process through volatile CPU registers, and are lost in the event of a crash failure. All the instructions discussed are commonly available in many modern processor architectures, such as Intel 64 [10] and AMD64 [11]. An RME algorithm may only use a subset of these instructions.

Procedure CAS(*address, old, new*)

75	**if** *old* = *address* **then**
76	*address* := *new*
77	**return** true
78	**else**
79	**return** false

Procedure FAS(*address, new*)

80	*old* := *address*
81	*address* := *new*
82	**return** *old*

Procedure FAA(*address, val*)

83	*old* := *address*
84	*address* := *address* + *val*
85	**return** *old*

Procedure FAI(*address*)

86	*old* := *address*
87	*address* := *address* + 1
88	**return** *old*

Fig. 4.15 Commonly supported atomic read-modify-write (RMW) instructions

References

1. Edsger W. Dijkstra. Solution of a problem in concurrent programming control. *Communications of the ACM (CACM)*, 8(9):569, 1965.
2. Wojciech Golab and Aditya Ramaraju. Recoverable mutual exclusion. In *Proc. of the 35th ACM Symposium on Principles of Distributed Computing (PODC)*, pages 65–74, 2016.
3. Wojciech Golab and Aditya Ramaraju. Recoverable mutual exclusion. *Distributed Computing (DC)*, 32(6):535–564, 2019.
4. Leslie Lamport. The mutual exclusion problem: part II – statement and solutions. *Journal of the ACM (JACM)*, 33(2):327–348, 1986.
5. Prasad Jayanti and Anup Joshi. Recoverable FCFS mutual exclusion with wait-free recovery. In *Proc. of the 31th International Symposium on Distributed Computing (DISC)*, pages 30:1–30:15, 2017.
6. Wojciech Golab and Danny Hendler. Recoverable mutual exclusion under system-wide failures. In *Proc. of the 37th ACM Symposium on Principles of Distributed Computing (PODC)*, pages 17–26, 2018.
7. Aditya Ramaraju. RGLock: Recoverable mutual exclusion for non-volatile main memory systems. Master's thesis, University of Waterloo, 2015.
8. Daniel Katzan and Adam Morrison. Recoverable, abortable, and adaptive mutual exclusion with sublogarithmic RMR complexity. In *Proc. of the 24th International Conference on Principles of Distributed Systems (OPODIS)*, pages 15:1–15:16, 2021.

9. James H. Anderson, Yong-Jik Kim, and Ted Herman. Shared-memory mutual exclusion: major research trends since 1986. *Distributed Computing (DC)*, 16(2-3):75–110, 2003.

10. Intel. *Intel 64 and IA-32 Architectures Software Developer's Manual, Volume 2A: Instruction Set Reference, A-M*, September 2016.

11. AMD. *AMD64 Architecture Programmer's Manual Volume 3: General Purpose and System Instructions*, September 2019.

Load and Store Based Algorithms

5

Although modern multiprocessors support a variety of powerful read-modify-write primitives, algorithms that use only atomic reads and writes (also known as loads and stores) are arguably more portable. RME algorithms in this category were first proposed by Golab and Ramaraju [1, 2], and serve as useful building blocks of more elaborate algorithms that leverage read-modify-write primitives, particularly by protecting the recovery section, to yield faster RME algorithms. One of the algorithms presented in this chapter solves the RME problem for n processes and provides $O\ (\log n)$ RMR complexity per passage, which is optimal because it matches the lower bound established by Attiya, Hendler, and Woelfel [3] for conventional mutual exclusion—a strictly easier problem.

5.1 Algorithm for Two-of-n-Processes

The first algorithm is obtained by transforming Yang and Anderson's two-process local-spin mutex [4]. Like its predecessor, this algorithm is designed for participation by two out of n processes at a time, meaning that at most two processes at a time execute a super-passage through the mutex. The execution paths of the two processes are distinguished by a special argument $Side \in \{\text{left}, \text{right}\}$ of the procedures Recover, Enter, and Exit, which is used internally as the index of an associative array (e.g., see line 89). The rules regarding the values of the argument are two-fold:

1. A process must use the same value of $Side$ consistently in each super-passage; and
2. If two processes are executing super-passages concurrently then they must use distinct values of $Side$.

© The Author(s), under exclusive license to Springer Nature Switzerland AG 2023
S. Dhoked et al., *Recoverable Mutual Exclusion*, Synthesis Lectures on
Distributed Computing Theory, https://doi.org/10.1007/978-3-031-20002-1_5

Shared variables:

- T: long process ID or $\bot$, initially $\bot$
- $C[\ldots]$: associative array indexed by left and right, each element a tuple of the form $\langle$long process ID or $\bot$, integer$\rangle$, all elements initially $\langle\bot, 0\rangle$
- $P[1\ldots n]$: array of integer spin variables, element $P[i]$ local to process p_i in the DSM model, all elements initially zero

Private variables:

- $rival$: integer, uninitialized

Definitions:

- $other(side) = \begin{cases} \text{left} & \text{if } side = \text{right} \\ \text{right} & \text{if } side = \text{left} \end{cases}$

Procedure Recover $(side)$ for process p_i

---

89 **if** $C[side] = \langle i, 1\rangle$ **then**
90      $\langle rival, \ldots\rangle := C[other(side)]$
91   **if** $rival \neq \bot$ **then**
92          $P[rival] := 2$

93 **else if** $C[side] = \langle\bot, 2\rangle$ **then**
94   execute lines 107 to 110 of Exit $(side)$

Procedure Enter $(side)$ for process p_i

---

95 $C[side] := \langle i, 1\rangle$
96 $T := i$
97 $P[i] := 0$
98 $\langle rival, \ldots\rangle := C[other(side)]$
99 **if** $rival \neq \bot$ **then**
100   **if** $T = i$ **then**
101   **if** $P[rival] = 0$ **then**
102             $P[rival] := 1$

103   **await** $P[i] \geq 1$
104   **if** $T = i$ **then**
105   **await** $P[i] = 2$

Procedure Exit $(side)$ for process p_i

---

106 $C[side] := \langle\bot, 2\rangle$
107 $rival := T$
108 **if** $rival \notin \{i, \bot\}$ **then**
109      $P[rival] := 2$
110 $C[side] := \langle\bot, 0\rangle$

Fig. 5.1 Recoverable extension of Yang and Anderson's two-process mutex

We will use the notation $Side_i$ in reference to the value of the argument $Side$ for process p_i that is executing a super-passage at the end of a given finite history. The shorthand notation other $Side$ will denote the "opposite side" to the one being used by a given process.

The recoverable two-process mutex for process p_i is presented in Fig. 5.1. There are three shared variables: (a) T is used to break symmetry at line 96, (b) C is an associative array used to track the progress of a process at lines 95, 106 and 110, and (c) P is an array of spin variables used for busy-waiting at lines 103 and 105. The entry and exit sections generally follow the structure of Yang and Anderson's algorithm, but there are important differences: (i) the elements of array C, which hold the IDs of processes competing for the CS in the original algorithm, are augmented with an integer tag that indicates progress through a passage; and (ii) an additional write to C is added at line 110 of Exit so that a recovering process can detect whether it crashed in the exit section. In the body of Enter, starting at line 95, a process proceeds directly to the CS unless it encounters a rival at line 99, in which case the busy-wait loop at line 103 ensures that both processes have progressed past line 96, and the second busy-wait loop at line 105 gives priority to the process that executed line 96 the earliest. Exit checks for a rival at line 107, and hands over the critical section at line 109, if required.

The recovery section at lines 89 to 94 is a new addition, as compared to Yang and Anderson's algorithm, and is executed whenever a process transitions out of the NCS. If process p_i is not in cleanup then $C[Side(i)] = \langle \perp, 0 \rangle$ holds while it executes Recover, and so the conditions tested at lines 89 to 93 are both false. In that case, p_i proceeds directly to Enter, which ensures bounded recovery. Otherwise, p_i tries to determine where it failed. Lines 89 and 92 handle a crash inside Enter (or the CS), in which case p_i determines at line 90 whether it has a rival p_j. If so, then p_i assigns $P[j] = 2$ at line 92 to ensure that the rival is not stuck at line 103 or line105, and then repeats the entry section. Lines 93 to 94 handle a crash inside Exit, in which case p_i repeats the body of the exit section and then proceeds to the entry section.

The main correctness properties of the two-of-n-process algorithm are stated below in Theorem 5.1.

Theorem 5.1 *The RME algorithm presented in Fig. 5.1 satisfies ME, JJ-SF, BE and BR. Furthermore, its worst-case RMR complexity per passage is $O(1)$ in the CC and DSM models.*

5.2   Adding Bounded Critical Section Reentry

Assuming independent process failures, a recoverable mutual exclusion algorithm can be augmented easily with the BCSR property using only $O(n)$ additional shared variables and $O(1)$ additional RMRs per passage it the CC and DSM models. Figure 5.2 presents a transformation that achieves this goal by tracking CS ownership using an array $C[1\ldots n]$ of bits. A process p_i assigns $C[i] := 1$ shortly before entering the CS at line 115, and resets this variable at line 116 shortly after clearing the CS. If p_i crashes in the CS, which implies $C[i] = 1$, then it bypasses the body of the recovery and entry sections in each subsequent passage until it re-enters the CS and reaches the exit section.

Shared variables:

- $mtxB$: base mutex, recoverable and provides CS continuity
- $C[1\ldots n]$: array of integer, all elements initially zero

---

Procedure Recover() for process p_i

---
111 **if** $C[i] = 0$ **then**
112 $\quad\lfloor\quad mtxB$.Recover()

Procedure Enter() for process p_i

---
113 **if** $C[i] = 0$ **then**
114 $\quad\mid\quad mtxB$.Enter()
115 $\quad\lfloor\quad C[i] := 1$

Procedure Exit() for process p_i

---
116 $C[i] := 0$
117 $mtxB$.Exit()

Fig. 5.2 Transformation from recoverable base mutex to recoverable target mutex that satisfies the BCSR property

The transformation assumes that the base mutex, denoted $mtxB$ in Fig. 5.2, satisfies the CS continuity property introduced earlier in Chap. 4, which allows a process that crashes in the CS to bypass the recovery and entry sections, proceeding directly to the CS upon recovering.[1] The two-of-n-process algorithm presented earlier in Sect. 5.1 meets this criterion, as does the n-process algorithm presented later on in Sect. 5.3. The BCSR transformation must be applied to the algorithm from Sect. 5.1 before it can be used as a building block of the algorithm in Sect. 5.3. The latter algorithm satisfies BCSR naturally, and does not require the transformation.

Theorem 5.2 *The transformation presented in Fig. 5.2 ensures BCSR, and preserves the following correctness properties of the base mutex $mtxB$: ME, DF, SF, JJ-SF, TE, BE, BR, k-BR, k-FCFS and CS continuity. Furthermore, the transformation also preserves its asymptotic worst-case RMR complexity per passage in the CC and DSM models.*

5.3   An Algorithm for n Processes

Golab and Ramaraju's n-process solution is modeled after Yang and Anderson's n-process algorithm [4], which is based on the arbitration tree of Kessels [5]. The algorithm is structured as a binary tree of height $O(\log n)$ where each node is a two-process mutex implemented

---

[1] This assumption is only compatible with the individual process failure model. Refer to footnote 5 in Chap. 4.

using the algorithm described in Sect. 5.1, and augmented for BCSR using the transformation described in Sect. 5.2. Each process is mapped statically to a leaf node in the tree as follows: process p_i enters at leaf node number $\lceil i/2 \rceil$ counting from 1. Furthermore, $Side = \mathsf{left}$ at leaf level if i is odd and $Side = \mathsf{right}$ if i is even.

Shared variables:

- a complete binary tree containing at least $\lceil n/2 \rceil$ and fewer than n leaf nodes, numbered starting at 1, and where each node is an instance of the algorithm from Sect 5.1 augmented with BCSR using the transformation from Sect 5.2 *root* denotes the root node of the tree

---

Procedure `Recover()` for process p_i

```
// no special recovery actions required
```

Procedure `Enter()` for process p_i

```
 1  node := leaf node ⌈i/2⌉
 2  if i is odd then
 3  │   side := left
 4  else
 5  │   side := right
 6  while node ≠ ⊥ do
 7  │   node.Recover(side)
 8  │   node.Enter(side)
 9  │   if node is the root then
10  │   │   node := ⊥
11  │   else if node is a left child then
12  │   │   side := left
13  │   │   node := parent of node
14  │   else
15  │   │   side := right
16  │   │   node := parent of node
```

Procedure `Exit()` for process p_i

```
17  node := ⊥
18  repeat
19  │   if node = ⊥ then
20  │   │   node := root
21  │   else
22  │   │   node := child of node on the path to leaf node ⌈i/2⌉
23  │   if node is a leaf and i is odd, or if leaf node ⌈i/2⌉ is in the left subtree of node then
24  │   │   side := left
25  │   else
26  │   │   side := right
27  │   node.Exit(side)
28  until node is a leaf
```

Fig. 5.3 Recoverable extension of Yang and Anderson's n-process mutex

The recoverable n-process mutex is presented in Fig. 5.3 for process p_i. Detailed pseudo-code is analogous to the non-recoverable n-process algorithm in [4] except that the two-process instances are implemented using our recoverable two-process mutex. The recovery section is empty since no additional recovery actions are required beyond those performed internally by the two-process mutex instances. The entry section entails executing the recovery and entry section (back to back) of each two-process mutex on the path from the designated leaf node of a process to the root, in that order. The exit section releases the two-process mutex instances in the opposite order, namely from root to leaf. The flow of control is analogous to the example of nested locks presented earlier in Fig. 4.6 of Chap. 4, and relies crucially on critical section re-entry.

Theorem 5.3 asserts the main correctness properties of the n-process algorithm.

Theorem 5.3 *The algorithm presented in Fig. 5.3 satisfies ME, JJ-SF, BCSR, BR, BE and CS continuity properties. Furthermore, its worst-case RMR complexity per passage is O (log n) in the CC and DSM models.*

References

1. Wojciech Golab and Aditya Ramaraju. Recoverable mutual exclusion. In *Proc. of the 35th ACM Symposium on Principles of Distributed Computing (PODC)*, pages 65–74, 2016.
2. Wojciech Golab and Aditya Ramaraju. Recoverable mutual exclusion. *Distributed Computing (DC)*, 32(6):535–564, 2019.
3. Hagit Attiya, Danny Hendler, and Philipp Woelfel. Tight RMR lower bounds for mutual exclusion and other problems. In *Proc. of the 40th ACM Symposium on Theory of Computing (STOC)*, pages 217–226, 2008.
4. Jae-Heog Yang and James H. Anderson. A fast, scalable mutual exclusion algorithm. *Distributed Computing (DC)*, 9(1):51–60, 1995.
5. Joep L. W. Kessels. Arbitration without common modifiable variables. *Acta Informatica*, 17:135–141, 1982.

Sublogarithmic Algorithms

6

To our knowledge, there are three RME algorithms whose worst-case RMR complexity for a passage is sub-logarithmic in the number of processes in the system, specifically $O\left(\frac{\log n}{\log \log n}\right)$ [1–3]. We consider only algorithms based on widely supported primitives in this chapter.

6.1 Golab and Hendler's Algorithm

Golab and Hendler [1] devised an algorithmic template, applicable to both the CC and DSM models, for constructing RME algorithms with sub-logarithmic RMR complexity. The main idea in this work is that an algorithm that guarantees $O(1)$ RMR complexity in the absence of failures and $O(n)$ RMRs in the worst case can be used, under certain conditions, as a building block of an arbitration tree that guarantees $O\left(\frac{\log n}{\log \log n}\right)$ RMR complexity in all passages. Assuming without loss of generality that $n = \Delta^{\Delta}$ for some positive integer Δ, this is accomplished using a Δ-ary tree of height Δ, with n leaf nodes. The leaf nodes serve as designated entry points into the arbitration tree, and the interior nodes are node-level RME locks accessed by Δ of n processes concurrently. The high-level tree structure is borrowed from Hendler and Woelfel's earlier work on randomized mutual exclusion [4], and is used to solve the RME problem deterministically without requiring intricate promotion mechanisms; the only way to enter the CS is to ascend from leaf level to the root node by capturing all the node-level locks along the way. The new insight due to Golab and Hendler is a precise characterization of correctness properties for the node-level lock that suffice for correctness of the arbitration tree.

The salient properties of the node-level RME lock are as follows. First, critical section re-entry (CSR) is necessary to correctly recover the nested node-level locks, as in Ramaraju and Golab's load-store algorithm described earlier in Chap. 5. Second, the node-level lock must be a *ported lock* in the sense that different subsets of n processes can access it at

S. Dhoked et al., *Recoverable Mutual Exclusion*, Synthesis Lectures on Distributed Computing Theory, https://doi.org/10.1007/978-3-031-20002-1_6

different times, subject to the constraint that only Δ processes are executing a super-passage concurrently. The port through which a process accesses the lock can be viewed as a temporary identity assigned to that process for the duration of a super-passage, and is determined uniquely by the child node of the arbitration tree from which the process ascended. In terms of RMR complexity, it is sufficient that the node-level lock has linear complexity in the worst case (i.e., $O(\Delta)$ RMRs with Δ ports) and $O(1)$ RMRs in any passage that is not 0-failure-concurrent. In other words, when a process p crashes while competing at a certain level ℓ in the tree, the failure can increase p's RMR complexity in its next passage but it does not affect the RMR complexity of passages executed by other processes that traverse level ℓ without crashing. The desired behaviour is illustrated in Fig. 4.4a of Chap. 4.

One of Golab and Ramaraju's constructions [5] comes close to achieving the above properties, yet it is not applicable because it guarantees $O(1)$ RMR complexity only in passages that are not 2-failure-concurrent (see Fig. 4.4b of Chap. 4). As a result, Golab and Hendler devised a Δ-ported RME lock with tighter RMR guarantees in the CC model (and unbounded RMR complexity in the DSM model), using as a starting point the well-known queue-based mutex lock of Mellor-Crummey and Scott [6]. We reproduce this algorithm, known commonly as the MCS lock, in Fig. 6.1 for completeness. Although the complicated recovery section of Golab and Hendler's queue lock was later found to be incorrect by Jayanti, Jayanti, and Joshi [2], its presentation in [1] leads to some important insights regarding the effect of process crashes on the queue structure, and techniques for repairing this structure during recovery.

We begin with an overview of the venerable MCS lock, which is presented in simplified[1] form in Fig. 6.1. The algorithm uses a queue of QNode structures, each one representing a distinct process, to track the order in which processes executed the algorithm's doorway (lines 146 to 147). This process queue is implemented using a singly-linked list, where pointers between nodes are directed from predecessor to successor.[2] In the absence of crash failures, process p_i appends its QNode $q[i]$ to the process queue by executing a non-atomic sequence of instructions: it updates the tail pointer T using FAS at line 147, saves the previous value of the tail pointer to the private variable $pred$, and finally establishes a pointer from the predecessor's node to $q[i]$ at lines 148 to 150 if a predecessor is discovered (i.e., $pred \neq$ **null**). The exit section begins with an attempt to resolve the successor node at lines 152 to 155, if it is not already known. The most distinctive feature of the MCS algorithm is that process p_i first tries to delete its node $q[i]$ from the process queue in this part of the exit section. This is accomplished by executing a CAS instruction at line 153 to reset T back to its initial value of **null**. On success, p_i is able to depart immediately at line 154 as the successor either does not exist, or has not yet completed line 147. Otherwise, p_i is assured

---

[1] This version of the MCS algorithm is context-free, meaning that the queue nodes are managed internally rather than being supplied by the application as arguments to the procedures Enter and Exit.

[2] This sentence refers to the *next* pointer embedded in the QNode structure, which is a shared variable. The MCS algorithm also uses a private *pred* pointer in the entry section.

Define QNode: structure { *next*: pointer to QNode, *locked*: boolean }

Shared variables:

- T: pointer to QNode, initially **null**
- $q[1...n]$: array of pointers to QNode, element $q[i]$ local to process p_i in DSM model, initially each element points to a separate QNode itself initialized as $\langle$**null**, false$\rangle$

Private variables:

- *pred*: pointer to QNode, uninitialized

Procedure Enter() for process p_i

---

146   $q[i].next :=$ **null**
      // append p_i's QNode to the tail of the process queue
147   $pred :=$ FAS$(\&T, q[i])$
      // check for predecessor
148   **if** $pred \neq$ **null then**
          // link with predecessor
149   $\quad$ $q[i].locked :=$ true
150   $\quad$ $pred.next := q[i]$
          // wait for predecessor to release lock
151   $\quad$ **await** $\neg q[i].locked$

Procedure Exit() for process p_i

      // check for successor
152   **if** $q[i].next =$ **null then**
          // no known successor, try to unlink from tail of process queue
153   $\quad$ **if** CAS$(\&T, q[i],$ **null**$)$ **then**
          // unlinked successfully, no successor
154   $\quad\quad$ **return**
          // wait for successor to link with p_i
155   $\quad$ **await** $q[i].next \neq$ **null**
      // transfer lock ownership to successor
156   $q[i].next.locked :=$ false

Fig. 6.1 Mellor-Crummey and Scott's queue-based ME algorithm

that a successor p_j does exist, and so p_i waits at line 155 for p_j to set $q[i].next$ by completing line 150.[3] Finally, p_i transfers ownership of the lock to p_j by setting $q[i].next.locked$ to false at line 156, which allows p_j to progress beyond line 151.

To make the MCS lock recoverable, Golab and Hendler modified its structure in two ways: (i) the variable *pred* is moved from the private state of a process into the QNode, which resides in shared memory; and (ii) each process additionally stores a pointer to its current node using a designated entry of a shared *announcement array*.[4] This additional state makes it possible to not only recover the address of the node a process was using at

---

[3] The busy-wait loop at line 155 implies that the MCS lock lacks wait-free exit, except in low contention scenarios where the CAS instruction at line 153 succeeds.

[4] A new QNode is allocated in each super-passage, and memory reclamation is not discussed.

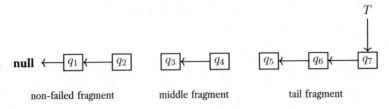

non-failed fragment   middle fragment   tail fragment

Fig. 6.2 Characterization of fragments in a broken process queue

the point of failure, but also to reconstruct portions of the process queue by following *pred* pointers when the *next* pointers are unavailable because line 150 was not reached. However, the queue cannot always be recovered in its entirety because the *pred* and *next* pointers of a process p_i's node cannot be updated atomically with p_i's execution of the FAS instruction in the entry section. As a result, a crash failure can occasionally split the process queue into multiple fragments.

Golab and Hendler analyzed different categories of fragments that may arise in a failure scenario. Formally, a fragment is a maximal sequence of nodes linked by *pred* pointers, meaning that node q_i is the immediate predecessor of node q_j if $q_j.pred = \&q_i$.[5] In the GH algorithm, a node is marked as *finished* in the last step of the exit section, and such nodes are excluded from fragments to ensure that each fragment contains at most n queue nodes, which is important for RMR complexity. Bearing this observation in mind, the first and last node of a fragment determine its category as follows:

1. a *non-failed* fragment starts with a node where *pred* is **null**, indicating the absence of a predecessor, or where *pred* points to a finished node
2. a *tail fragment* includes the tail node identified by the tail pointer T, and may itself be a non-failed fragment
3. a *middle fragment* is any fragment that does not meet the criteria of a tail fragment or non-failed fragment

Examples of different fragments are illustrated in Fig. 6.2.

To recover the process queue, Golab and Hendler proposed a conceptually simple but difficult to implement strategy. A recovering process p_i first identifies any existing queue fragments, determines which fragment contains its node, and in some cases attaches its fragment to another judiciously chosen fragment. The queue structure is analyzed by walking *pred* pointers backward from the nodes stored in the announcement array, as well as from T. The analysis occurs only when p_i's *pred* pointer is unknown, meaning that it holds

---

[5] We refer in this statement to the GH algorithm, where *pred* is a shared variable. With respect to the MCS lock in Fig. 6.1, the corresponding condition is that $q_i.next = \&q_j$ or p_j is between line 147 and line 150 with $pred = \&q_i$.

a special initial value that is different from the initial (**null**) value of T.[6] This condition implies immediately that p_i's node is the head of its corresponding fragment, and hence the fragment is either a middle fragment, or a tail fragment other than a non-failed fragment. Any process queued behind p_i in this fragment is stuck until the process queue is repaired.

The mechanism of repair depends on the size and type of p_i's fragment. If p_i's fragment has size one and is not the tail fragment, then there is no danger of another process waiting behind p_i, and p_i can leave the recovery section to begin a new passage. Alternatively, if p_i's fragment is the only fragment, then p_i's node is at the head of the process queue and p_i proceeds to the CS with *pred* set to **null**. Finally, the most challenging case occurs when there are multiple fragments, and p_i must connect its own fragment to another one. In general, there are two ways to form the connection: (i) if p_i's fragment f is not the tail fragment, then it can be appended behind the tail fragment by swapping the tail node of f into T using FAS, and then setting p_i's *pred* pointer to the value returned by the FAS instruction; and (ii) p_i's fragment f can be appended behind a non-tail fragment f' by setting p_i's *pred* pointer to the address of the last node in f'. Both strategies are useful in implementing a recoverable MCS lock.

The simplified overview of the GH algorithm given so far omits a number of subtle but important technical details. On closer inspection, the reconstruction of the process queue is an endeavour fraught with peril due to the assumption of independent process failures, which implies that multiple processes may recover concurrently, and that recovery actions by one process may be concurrent with failure-free passages of the other processes. Next, we identify some specific technical problems that complicate recovery, and describe informally the corresponding solutions:

Problem P1: Processes may attempt to reconstruct the process queue concurrently in the recovery section.

Solution S1: Recovery actions are protected using an *auxiliary lock*, for example Golab and Ramaraju's n-process tournament tree [7] (see Chap. 5).

Problem P2: Queue fragments may coalesce spontaneously as processes set their *pred* pointers in the entry section.

Solution S2: While analyzing the queue structure, a recovering process that encounters a node where *pred* is set to its initial value ($\perp$ in the GH algorithm) waits for the pointer to settle (i.e., take on a different value). The wait loop terminates when *pred* is **null** or points to another node.

Problem P3: The combination of S1 and S2 can lead to deadlock if a recovering process p_i waits for the *pred* pointer of another process p_j to settle while p_j crashes and then attempts to acquire the auxiliary lock (held by p_i) during recovery.

---

[6] The initial value of *pred* must be different from the initial value of *tail* (i.e., **null**) so that a recovering process can distinguish between the situation where it never updated *pred*, and the situation where it did update *pred* with a **null** value, indicating that it was the head of the process queue when it executed the FAS instruction to update *tail*.

Solution S3: A recovering process p_j whose *pred* pointer still holds the initial value sets *pred* to a special address (e.g., of a sentinel node) indicating that it has failed and is recovering. A process p_i that holds the auxiliary lock described in S1 breaks out of the busy-wait loop described in S2 when p_j's *pred* pointer settles on the special address.

Problem P4: As new nodes may be appended to the process queue during recovery, the tail fragment may grow, a new fragments may also appear.

Solution S4: The main concern is that a recovering process p_i will incorrectly categorize a fragment f, and use an unsafe strategy to append its own fragment behind f. For example, if p_i categorizes f as a tail fragment, but a new tail fragment f' develops while p_i is applying strategy (i), then p_i may end up appending its own fragment to f' instead of f. This particular anomaly is benign because strategy (ii) will update the tail pointer correctly and establish p_i's *pred* pointer. However, if p_i categorizes f as a middle fragment when it is in fact the tail fragment, then applying strategy (ii) can lead to multiple nodes sharing the same predecessor, and a violation of starvation freedom. To rule out this case, the queue structure is analyzed by first following pointers from entries of the announcement array, and then following the tail pointer T as the last order of business. Any additional fragments created by a moving tail pointer are then excluded from the analysis, and cannot be confused for middle fragments.

Problem P5: When the recovering process p_i connects its fragment to another fragment f, p_i must wait for its new predecessor to transfer ownership of the lock.

Solution S5: Process p_i executes a portion of the entry section where it sets the predecessor's *next* pointer and awaits for the predecessor to unlock p_i's node.

Problem P6: When the recovering process p_i connects its fragment to another fragment f, its new predecessor may have already deleted itself from the process queue by completing a successful CAS instruction in the exit section.

Solution S6: The main concern is that the predecessor never transfers ownership of the lock to p_i. This problem arises only when f is a non-failed fragment, and so p_i first attempts to avoid the problem by ensuring that f is either a middle fragment, or the tail fragment.[7] If this is not possible because p_i's fragment is the tail fragment and there are no middle fragments to append to, then there exists at least one non-failed fragment, as otherwise p_i's fragment is the only one. This final case is the most challenging one since there can be multiple non-failed fragments,[8] and p_i cannot reliably determine their chronological

---

[7] In this case, choosing f to be the tail fragment ensures that f is a stuck tail because the tail cannot be non-failed if there exists at least one other fragment (i.e., p_i's).

[8] This property arises from the MCS lock's unique exit protocol, where a process may delete its node from the queue and restore the tail pointer to its initial value of **null**. Because of this, multiple non-failed fragments may arise even in the absence of crash failures.

order. Even if it could identify the latest non-failed fragment, problem P6 could still arise. Rather than connecting to one of these fragments, p_i updates the *next* pointer of the last node in each fragment to ensure that its owner p_j can complete the exit section, and then waits for p_j to mark this node as finished. Once all the non-failed fragments have been discharged, p_i's fragment is either the only remaining fragment, or the tail has moved and p_i is now part of a middle fragment. Process p_i can proceed to the CS in the first case, and can append its fragment behind the tail in the second case.

The many technicalities involved in repairing the MCS queue were not fully addressed in the GH algorithm [1], and two specific issues were documented by Jayanti, Jayanti, and Joshi in [2]: deadlock is possible during analysis of the process queue due to a flaw in the implementation of Solution S3, and starvation is possible when queue fragments are improperly connected (i.e., tail fragment mistaken for a middle fragment) since Solution S4 was not implemented. The GH algorithm also does not explicitly address Problem P6. To our knowledge, Jayanti, Jayanti, and Joshi's (JJJ) algorithm avoids these pitfalls, partly thanks to more careful analysis, and also partly due to a wait-free exit section that circumvents P6 entirely. As explained in the next section, the JJJ algorithm confirms the RMR complexity bound claimed for the CC model by Golab and Hendler in [1], and also extends this bound to the DSM model.

6.2   Jayanti, Jayanti and Joshi's Algorithm

Jayanti, Jayanti and Joshi's sub-logarithmic RME algorithm [2] has the same high level structure as Golab and Hendler's sub-logarithmic RME algorithm. Specifically, it uses the same arbitration tree technique to progressively narrow the subset of processes competing to acquire the target lock until there is only one process left. Further, each node of the tree is a multi-ported RME lock with the same RMR complexity behavior. As in the GH algorithm, the number of ports of the multi-ported RME lock is set to $\Delta = \lceil \frac{\log n}{\log \log n} \rceil$ to obtain the desired sub-logarithmic RMR complexity.

The key difference is that Jayanti, Jayanti and Joshi's algorithm uses an alternative implementation of the multi-ported RME lock at each tree node that not only removes the race conditions present in Golab and Hendler's implementation that made it vulnerable to starvation, but also satisfies the following additional properties. First, it has sub-logarithmic RMR complexity for both CC and DSM models. Second, it only uses the FAS read-modify-write instruction. The GH implementation, on the other hand, has unbounded RMR complexity for the DSM model and also uses the CAS read-modify-write instruction.

To avoid repetition, in this section, we only describe the multi-ported RME lock used in JJJ's algorithm at each tree node assuming Δ ports. The lock uses three building blocks. The first building block is a Signal object, which is used by a process to *wait* until another process has reached a certain point in its execution. The second building block consists of a modified version of the well-known queue-based MCS lock, which is used to *serialize* critical section executions. Finally, the third building block is a repair procedure, which is used to *fix* the queue of the MCS lock that may have become fragmented due to a process crash.

6.2.1 Signal Object

A Signal object is accessed using two procedures, namely Set and Wait. The object has a boolean value and is owned by a process. The Set procedure can only be invoked by its owner and the Wait procedure can only be invoked by any non-owner process. It is assumed that no two invocations of the Wait procedure overlap.

CC Model
It is easy to implement the object in the CC model such that both procedures have $O(1)$ RMR complexity per invocation. In the Set procedure, the owner process sets the value of the object to true. In the Wait procedure, a non-owner process spins until the value of the object becomes true.

DSM Model
To achieve $O(1)$ RMR complexity in the DSM model, a non-owner process has to spin on its local variable when waiting. To that end, a non-owner process, upon invoking the Wait procedure, first registers the address of its spin location with the object and then reads the value of the object. If it finds the value to be false, then it spins on the location it has registered. The owner process, upon invoking the Set procedure, first sets the value of the object to true and then checks if an address of a spin location has been registered with the object. If yes, then it proceeds to release the process spinning on that location. Pseudocode for the DSM algorithm is given in Fig. 6.3.

Theorem 6.1 asserts the RMR complexity behavior of the algorithm in Fig. 6.3.

Theorem 6.1 *The implementation of Signal object in Fig. 6.3 has $O(1)$ RMR complexity in the DSM model.*

6.2.2 Modified MCS Lock

As in GH's algorithm, an announcement array, with one entry for each port, is used to store the address of the queue node being used to gain entry into critical section through a port.

Shared variables:

- *Bit*: boolean object, initially false
- *GoAddr*: pointer to boolean object, initially **null**

Private variables:

- *addr*, *go*: pointer to boolean object, uninitialized

Procedure Set() for process p_i

---

157   Bit := false
158   $addr$:= $GoAddr$
159   **if** $addr \neq$ **null then**
160 | $*addr$:= true

Procedure Wait() for process p_i

---

161   go := pointer to a new boolean object
162   $*go$:= false
163   $GoAddr$:= go
164   **if not** *(Bit)* **then**
165 | **await** $(*go)$

Fig. 6.3 An implementation of Signal object in the DSM model

It allows other processes competing for the same lock to know which queue nodes are still in use, which we refer to as *active* nodes. The announcement array is denoted by *Node*.

JJJ's algorithm modifies the queue node of the MCS lock to contain three fields: Pred, NonNullPred and CSComplete. The first field Pred is used to store the address of the predecessor node once the node has been appended to the queue. The other two fields store a Signal object described earlier, which are used by the owner of the node to announce that it has reached certain points in its execution. The first Signal object stored in NonNullPred field is used to inform a repairing process that Pred field now contains a non-null value. The second Signal object stored in CSComplete field is used to inform the owner of the successor node that it can now enter its critical section.

Assume, for now, that a process does not fail while trying to acquire the (modified) MCS lock. The steps for acquiring the lock are as follows. The process first creates a new node and writes its address to the entry in *Node* array corresponding to the port through which it is trying to acquire the lock. The process then appends its node to the queue by executing the FAS instruction on the tail pointer, which also returns the address of its predecessor node. It next writes this address to the Pred field of its own node. It then sets NonNullPred field of its own node, which is a Signal object, to true. This releases any repairing process that may be waiting for Pred field of the node to attain a non-null value. It finally waits until CSComplete field of its predecessor node, which is also a Signal object, has been set to true, indicating that it now has the lock. Note that, unlike in the original MCS lock, a node is never removed from the queue. This helps to ensure that the tail pointer always has a non-null value. Initially, this is achieved by initializing the tail pointer to point to a special sentinel node, denoted by SpecialNode. Intuitively, this sentinel node represents the *beginning* of the queue.

If a process fails after performing the FAS instruction but before it could write its return value to the Pred field of its node, then there is no easy way to recover the address of the predecessor node in general and the queue may become fragmented. The crashed process, upon recovery, is responsible for attaching the fragment containing its node to the queue by executing a repair procedure. The procedure guarantees that, upon completion, the node of the crashed process has a proper predecessor node. As before, the process then waits until CSComplete field of its predecessor node has been set to true indicating that it now has the lock.

Upon leaving its critical section, the process sets CSComplete Signal object of its node to true and clears the entry for the port it used in the *Node* array. The former step releases the lock and the latter step indicates that the node has been retired and is no longer in use.

6.2.3   Repair Procedure

As mentioned in Sect. 6.1, the repair procedure is fraught with peril. To eliminate many of the race conditions, execution of the repair procedure by multiple processes is serialized using a possibly inefficient Δ-ported auxiliary RME lock, similarly to the GH algorithm. (Recall that the node-level lock is also a Δ-ported RME lock.) This lock, denoted by RLock, is assumed to have *linear* worst-case RMR complexity. The repair procedure itself has linear RMR complexity and is executed within the critical section of RLock.

To aid in repair, JJJ's algorithm uses the Pred field of a node to provide information about the owner (of the node) to other processes in certain situations. To that end, the algorithm creates three sentinel nodes specifically used to indicate status, namely Crash, InCS and Exit. The first sentinel node Crash is used to indicate that the owner has lost the address of the predecessor node obtained by executing the FAS instruction on the tail pointer due to a crash and, thus, its node is not yet attached to the queue. The second sentinel node InCS is used to indicate that the owner is currently executing its critical section. The third sentinel node Exit is used to indicate that the owner has completed its critical section.

Now, to execute the repair procedure, a process first acquires RLock. It next scans the *Node* array to identify all active nodes. For every active node found, it waits until the Pred field of that node has attained a non-null value indicating that an attempt has been made to append that node to the queue, which may or may not have succeeded. The process then constructs a directed graph whose set of vertices is given by the set of these active nodes along with their predecessors (provided the predecessor node is not a sentinel node used to indicate status), and whose set of edges is given by the predecessor relation. The process then computes the set of *maximal* paths of this graph, referred to as *fragments*. It next identifies three specific fragments as follows:

(a) *self-fragment:* it is the fragment that contains the node owned by the repairing process.
(b) *tail-fragment:* it is the fragment that contains the node referred to by the tail pointer.
(c) *head-fragment:* it is the fragment that contains at least one node that is either currently in or enabled to enter its critical section.

Note that the self-fragment is guaranteed to exist. However, tail and head fragments may not exist. For example, the tail-fragment may not exist if the node referred to by the tail pointer is no longer in use (i.e., has been retired). Likewise, the head-fragment may not exist if all active nodes found are owned by processes that crashed immediately after performing their respective FAS instructions. Further, note that the self and tail fragments may be the same. Also, tail and head fragments may be the same. To fix the queue, the repairing process then either

(a) appends the self-fragment after the tail node, or
(b) attaches the self-fragment to the end of the head-fragment, or
(c) promotes the self-fragment to become the new head-fragment.

The first case applies if either the tail-fragment does not exist or the tail-fragment contains a node whose owner has already completed its critical section. The second case applies if the first case does not apply and the head-fragment exists. Finally, the third case applies if none of the above two cases apply. In the first case, the process moves the tail pointer to the last node of the self-fragment. In the second case, it sets the Pred field of its own node to the last node of the head fragment. Finally, in the last case, it sets the Pred field of its own node to the special sentinel node (SpecialNode).

6.2.4 Formal Description

The pseudocode of the Δ-ported RME lock for a tree node is given in Figs. 6.4, 6.5 and 6.6. The algorithm has been modified from its original version to adhere to the Golab and Rama-raju's execution model in two ways. In every passage, a process executes the Recover, Enter and Exit procedures in order (with the critical section in between Enter and Exit). Second, there are no "jump" or "go to" statements from one procedure to another, which are not allowed in a high-level programming language such as C/C++.

Figure 6.4 shows the data structures and shared variables used in the algorithm. Figure 6.5 shows the pseudocode for the Recover, Enter and Exit procedures. Figure 6.6 contains the pseudocode for the Repair procedure, which is used by the Recover procedure to fix the queue as needed. All four procedures take the port number that the invoking process is using to acquire the lock as an argument.

In the Recover procedure, a process checks if it has an active node (line 167) whose Pred field is null (lines 168 to 169). If so, it updates the Pred field of its node to indicate that

Type definitions:

> **struct** QNode {
>   Pred: pointer to QNode
>   NonNullPred: Signal object described in section 6.2.1
>   CSComplete: Signal object described in section 6.2.1
> }

Shared variables:

- Crash: sentinel QNode, initially { &Crash, true, false }
- InCS: sentinel QNode, initially { &InCS, true, false }
- Exit: sentinel QNode, initially { &Exit, true, false }
- SpecialNode: sentinel QNode, initially { &SpecialNode, true, true }
- $Node[1\ldots\Delta]$: array of pointers to QNode, all elements initially **null**
- Tail: pointer to QNode, initially &SpecialNode
- RLock: a Δ-ported RME lock that incurs $O(\Delta)$ RMRs per passage in both CC and DSM models

Private variables:

- $mynode, mypred, tail, cur, curpred$: pointer to QNode, uninitialized
- V: set of pointers to QNode, uninitialized
- E: set of tuples, each tuple of the form ⟨pointer to QNode, pointer to QNode⟩, uninitialized
- $tailpath, headpath, mypath$: sequence of pointers to QNode, uninitialized
- $Paths$: set of sequences of pointers to QNode, uninitialized

Fig. 6.4 The Δ-ported recoverable lock used in Jayanti, Jayanti and Joshi's sub-logarithmic RME algorithm. Presentation continued in Figs. 6.5 and 6.6

the node is not "attached" to the queue (line 170). It then notifies other processes that the Pred field of its node has a non-null value (line 171). It then examines its status to check whether the queue needs fixing (line 172) or it needs to complete its Exit section (line 177). In the first case, it invokes the repair procedure within a critical section (lines 173 to 176). In the second case, it invokes the Exit procedure (line 178).

In the Enter procedure, a process checks whether it already has an active node (line 179). If not, it allocates a new queue node (line 180) and attempts to append the node to the queue using the FAS instruction (line 182). It then writes the return value of the instruction to the Pred field of its node (line 183), which makes it persistent, and notifies other processes that the field now contains a non-null value (line 184). If it already has an active node, then it reads its address and that of its predecessor into its local variables (lines 185 to 186). If it does not have the lock already (line 187), then it waits for its predecessor to release the lock (line 188) and updates its status (line 189) indicating that it now holds the lock.

In the Exit procedure, a process updates its status (line 191), releases the lock (line 192) and retires its node (line 193).

In the Repair procedure, a process first checks whether it still needs to fix the queue (lines 194 to 196). If so, it scans the *Node* array to locate all active nodes and their predecessors (lines 202 to 211). It then constructs a directed graph induced by the predecessor relation (line 212), and identifies the three specific fragments mentioned earlier (lines 213 to

---

Procedure Recover(k) for process p_i

---

166 $mynode := Node[k]$
167 **if** $mynode \neq$ **null then**
168      $mypred := mynode$.Pred
169   **if** $mypred =$ **null then**
170          $mynode$.Pred:= &Crash
171      $mynode$.NonNullPred.Set()
172   **if** $mypred =$ &Crash **then**
173   RLock.Recover()
174   RLock.Enter()
175   Repair()
176   RLock.Exit()
177   **else if** $mypred =$ &Exit **then**
178   Exit()

Procedure Enter(k) for process p_i

---

179 **if** $Node[k] =$ **null then**
180      $mynode :=$ a new QNode
181      $Node[k] := mynode$
182      $mypred :=$ FAS(Tail, $mynode$)
183      $mynode$.Pred:= $mypred$
184      $mynode$.NonNullPred.Set()
185 $mynode := Node[k]$
186 $mypred := mynode$.Pred
187 **if** $mypred \neq$ &InCS **then**
188      $mypred$.CSComplete.Wait()
189      $mynode$.Pred:= &InCS

Procedure Exit(k) for process p_i

---

190 $mynode := Node[k]$
191 $mynode$.Pred:= &Exit
192 $mynode$.CSComplete.Set()
193 $Node[k] :=$ **null**

Fig. 6.5 The Δ-ported n-process recoverable lock used in Jayanti, Jayanti and Joshi's sub-logarithmic RME algorithm. Presentation continued from Fig. 6.4 and continued in Fig. 6.6

219). Next, it attaches the self-fragment to an appropriate location in the queue as explained earlier (lines 220 to 226).

Theorem 6.2 asserts the correctness properties of the Δ-ported n-process RME algorithm.

Theorem 6.2 *The Δ-ported n-process RME algorithm described in Figs. 6.4, 6.5 and 6.6 satisfies ME, JJ-SF, BCSR, BE, 0-BR and 1-FCFS properties.*

Theorem 6.3 asserts the complexity behavior of the Δ-ported n-process RME algorithm.

Procedure Repair(k) for process p_i

---

194   $mynode := Node[k]$
195   $mypred := mynode$.Pred
196   **if** $mypred \neq$ &Crash **then return**

197   $tail :=$ Tail
198   $V := \emptyset$
199   $E := \emptyset$
200   $tailpath := \bot$
201   $headpath := \bot$

202   **for** $j = 0$ **to** $\Delta - 1$ **do**
203 | $cur := Node[j]$
204 | **if** $cur =$ **null then continue**
205 | cur.NonNullPred.Wait()
206 | $curpred := cur$.Pred

207 | **if** $curpred \in \{$&Crash, &InCS, &Exit$\}$ **then**
208 | | $V := V \cup \{cur\}$
209 | **else**
210 | | $V := V \cup \{cur, curpred\}$
211 | | $E := E \cup \{(cur, curpred)\}$

212   Compute the set $Paths$ of maximal paths in the graph (V, E)

213   Let $mypath$ be the unique path in $Paths$ that contains $mynode$
214   **if** $tail \in V$ **then**
215 | let $tailpath$ is the unique path in $Paths$ that contains $tail$

216   **foreach** $\sigma \in Paths$ **do**
217 | **if** end(σ).Pred $\in \{$&InCS, &Exit$\}$ **then**
218 | | **if** start(σ).Pred $\neq$ &Exit **then**
219 | | | $headpath := \sigma$

220   **if** $tailpath = \bot \vee$ end($tailpath$) $\in \{$&InCS, &Exit$\}$ **then**
221 | $mypred :=$ FAS(Tail, start($mypath$))
222   **else if** $headpath \neq \bot$ **then**
223 | $mypred :=$ start($headpath$)
224   **else**
225 | $mypred :=$ &SpecialNode

226   $mynode$.Pred$:= mypred$

Fig. 6.6 The Δ-ported n-process recoverable lock used in Jayanti, Jayanti and Joshi's sub-logarithmic RME algorithm. Presentation continued from Fig. 6.5

Theorem 6.3 *The Δ-ported n-process RME algorithm described in Figs. 6.4, 6.5 and 6.6 has worst-case RMR complexity of $O(1)$ per passage if it is not 0-failure-concurrent and $O(\Delta)$ per passage otherwise in the CC and DSM models. Furthermore, it has RMR complexity of $O(1)$ for critical section reentry in a passage, when applicable, in the CC and DSM models.*

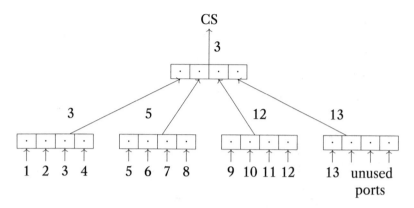

Fig. 6.7 Example of a tournament tree with $n = 13$ and $\Delta = 4$

6.3   Putting It All Together

As explained by Golab and Hendler [1] (see Sect. 6.1), a Δ-ported n-process RME lock like the one described in Figs. 6.4, 6.5 and 6.6 can be used as a building block of a Δ-ary tournament tree to obtain an RME lock with sub-logarithmic RMR complexity by setting $\Delta = \left\lceil \frac{\log n}{\log \log n} \right\rceil$. An illustrative example of a tournament tree is presented in Fig. 6.7. Pseudocode of a tournament tree-based RME algorithm is given in Fig. 6.8.

Theorem 6.4 asserts the correctness properties of Golab and Hendler's arbitration tree, which can be viewed as a generic transformation that improves the worst-case RMR complexity of an RME lock.

Theorem 6.4 *The RME algorithm described in Fig. 6.8 preserves the ME, JJ-SF, BCSR and BR properties of the Δ-ported n-process RME lock used as building block in the tournament tree.*

Theorem 6.5 asserts the RME complexity behaviour of Jayanti, Jayanti and Joshi's sublogarithmic RME algorithm [2], which is obtained by composing Jayanti, Jayanti and Joshi's queue-based RME algorithm (linear worst-case RMR complexity) with Golab and Hendler's arbitration tree.

Theorem 6.5 *The RME algorithm described in Fig. 6.8 has $O\left(\log n / \log \log n\right)$ RMR complexity when $\Delta = \left\lceil \frac{\log n}{\log \log n} \right\rceil$ in the CC and DSM models.*

Shared variables:

- a Δ-ary tree with $\Delta \geq 2$ that contains at least $\lceil \frac{n}{\Delta} \rceil$ leaf nodes and whose every node is a Δ-ported RME lock described in figs. 6.4 to 6.6, *root* denotes the root node of the tree
- *leaf*$[1 \ldots n]$: array of pointers to tree nodes, each element i points to the bottommost (first) tree lock to be acquired by process p_i

---

Procedure `Recover()` for process p_i

    `// no specific action required`

Procedure `Enter()` for process p_i

227 `EnterHelp(`*leaf*$[i]$, $\lfloor \frac{i-1}{\Delta} \rfloor$, $(i-1) \mod \Delta + 1)$

Procedure `Exit()` for process p_i

228 `ExitHelp(`*leaf*$[i]$, $\lfloor \frac{i-1}{\Delta} \rfloor$, $(i-1) \mod \Delta + 1)$

Procedure `EnterHelp(`*node*, *id*, *port*`)` for process p_i

229 *node*.`Recover(`*port*`)`
230 *node*.`Enter(`*port*`)`
231 **if** *node* $\neq$ *root* **then**
232   ⌊ `EnterHelp(`*node*.*parent*, $\lfloor \frac{id}{\Delta} \rfloor$, *id* $\mod \Delta + 1)$

Procedure `ExitHelp(`*node*, *id*, *port*`)` for process p_i

233 **if** *node* $\neq$ *root* **then**
234   ⌊ `ExitHelp(`*node*.*parent*, $\lfloor \frac{id}{\Delta} \rfloor$, *id* $\mod \Delta + 1)$

235 *node*.`Exit(`*port*`)`

Fig. 6.8 Golab and Hendler's tournament tree from [1]

References

1. Wojciech Golab and Danny Hendler. Recoverable mutual exclusion in sub-logarithmic time. In *Proc. of the 36th ACM Symposium on Principles of Distributed Computing (PODC)*, pages 211–220, 2017.
2. Prasad Jayanti, Siddhartha Jayanti, and Anup Joshi. A recoverable mutex algorithm with sub-logarithmic RMR on both CC and DSM. In *Proc. of the 38th ACM Symposium on Principles of Distributed Computing (PODC)*, pages 177–186, 2019.
3. Daniel Katzan and Adam Morrison. Recoverable, abortable, and adaptive mutual exclusion with sublogarithmic RMR complexity. In *Proc. of the 24th International Conference on Principles of Distributed Systems (OPODIS)*, pages 15:1–15:16, 2021.
4. Danny Hendler and Philipp Woelfel. Randomized mutual exclusion with sub-logarithmic rmr-complexity. *Distributed Computing (DC)*, 24(1):3–19, 2011.
5. Wojciech Golab and Aditya Ramaraju. Recoverable mutual exclusion. *Distributed Computing (DC)*, 32(6):535–564, 2019.

6. John M. Mellor-Crummey and Michael L. Scott. Algorithms for scalable synchronization on shared-memory multiprocessors. *ACM Transactions on Computer Systems (TOCS)*, 9(1):21–65, 1991.

7. Wojciech Golab and Aditya Ramaraju. Recoverable mutual exclusion. In *Proc. of the 35th ACM Symposium on Principles of Distributed Computing (PODC)*, pages 65–74, 2016.

Adaptive Algorithms

7

So far, we have measured the performance of an RME algorithm using its worst case RMR complexity. However, on average, an RME algorithm may incur lower RMRs in the absence of failures and/or contention. This class of algorithms is known as adaptive algorithms. In this chapter, we discuss adaptive algorithms whose RMR complexity varies with respect to the total number of processes in the system, and also with respect to the number of failures in the system.

7.1 Classification of RME Algorithms

Different failure scenarios warrant different RME algorithms. Dhoked and Mittal [1] classify RME algorithms by analyzing the RMR complexity of an RME algorithm (a) in the absence of failures (failure free RMR complexity), (b) in the presence of F failures (limited failures RMR complexity), and (c) in the presence of an unbounded number of failures (arbitrary failures RMR complexity).

Let $g(n, F)$ be a function of n and F, where $n \geq 1$ represents the number of processes in the system and $F \geq 0$ represents the number of failures that have occurred so far. Assume that $g(n, F)$ is a monotonically non-decreasing function of n and F, since this function is later used to represent the worst-case RMR complexity of an RME algorithm. The following performance measures are defined for the function $g(n, F)$.

PM 1. (Constantness) In the absence of failures, the function has a constant value independent of n. Formally, $g(n, 0) = O(1)$.

PM 2. (Adaptiveness) In order to capture adaptiveness, we define a function $\Delta(n)$ that captures the number of values of F for which the function changes.

© The Author(s), under exclusive license to Springer Nature Switzerland AG 2023
S. Dhoked et al., *Recoverable Mutual Exclusion*, Synthesis Lectures on
Distributed Computing Theory, https://doi.org/10.1007/978-3-031-20002-1_7

$$\Delta(n) = |\{F \mid g(n, F) < g(n, F + 1)\}|$$

With limited number of failures, there are three different cases.

(a) The function $g(n, F)$ has a trivial dependence on F. Formally, $\Delta(n) = \Omega(1)$.
(b) The function $g(n, F)$ has a strong dependence on F. Formally, $\Delta(n) = \omega(1)$.
(c) The function $g(n, F)$ has strong and sub-linear dependence on F. Formally, $\Delta(n) = \omega(1)$ and $g(n, F) = o(F)$.

PM 3. (Boundedness) With arbitrarily large number of failures, we identify two different cases:

(a) The function is finite-valued even as F tends to infinity. Formally, $\lim_{F \to \infty} g(n, F)$ is finite-valued.
(b) The function is bounded by a sub-logarithmic function of n. Formally, $\forall F : g(n, F) = O(\log n / \log \log n)$.

Figure 7.1 shows how the RMR complexity of algorithms increases with number of failures. Figure 7.2 shows a comparison of RMR complexity of algorithms based on boundedness as the number of failures increase.

Note that PM 2(b) implies PM 2(a), PM 2(c) implies PM 2(b) and PM 3(b) implies PM 3(a).

Consider an RME algorithm $\mathcal{A}$ and let $g(n, F)$ denote the *best known* bound on the worst-case RMR complexity of $\mathcal{A}$. Based on the performance measures satisfied by $g(n, F)$ $\mathcal{A}$ is labeled as follows:

1. Based on adaptiveness

   • *non-adaptive* if it does not satisfy PM 2(a).
   • *semi-adaptive* if it satisfies PM 2(a).
   • *adaptive* if it satisfies PM 2(b) (hence also PM 2(a)).
   • *super-adaptive* if it satisfies PM 2(c) (hence also PM 2(b) and PM 2(a)).

2. Based on boundedness

   • *unbounded* if it does not satisfy PM 3(a).
   • *bounded* if it satisfies PM 3(a).
   • *well-bounded* if it satisfies PM 3(b).

Fig. 7.1 RMR complexity of
adaptive algorithms

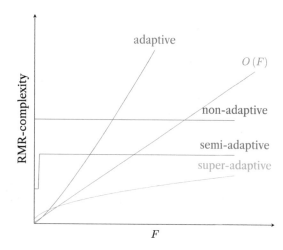

Fig. 7.2 RMR complexity of
bounded algorithms

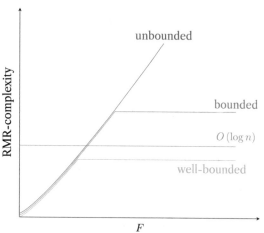

The rest of this chapter presents different types of adaptive algorithms. Golab and Ramaraju [2, 3] devise a transformation using which any (traditional) ME algorithm can be con-
verted into an adaptive RME algorithm. However, the algorithm as a result of this transformation is unbounded in the worst-case. Further, in order to bound the worst-case RMR
complexity, they devise a framework that transforms their unbounded adaptive RME algorithm into a bounded semi-adaptive one. Dhoked and Mittal [1] devise a transformation to
transform any RME algorithm into a super-adaptive one. They do this in three steps. First,
they define the problem of weakly RME and devise an optimal algorithm for the same. Second, they design a transformation built upon Golab and Ramaraju's framework to transform
any RME algorithm into a semi-adaptive one. Finally, the semi-adaptive algorithm is used
recursively to construct a super-adaptive transformation.

7.2   Adding Recoverability

This algorithm is used as a general transformation of traditional mutex algorithms (base lock) into recoverable ones (target lock). It is agnostic with regard to the internal structure of the base lock in that it deals with failures by resetting the entire base lock to its initial state, rather than by attempting delicate repairs. The algorithm assumes that the base lock provides a separate procedure Reset that can be executed by any process as long as no other process is accessing the base lock concurrently.

Pseudocode of the transformation for adding recoverability is given in Figs. 7.3 and 7.4.

Let $mtxB$ be the base lock. To ensure exclusive access for $mtxB$.Reset, an auxiliary recoverable lock, $mtxA$ is utilized by processes recovering from a failure. This lock ($mtxA$)

Shared variables:

- $mtxA$: auxiliary mutex, recoverable
- $mtxB$: base mutex
- $C[1 \ldots n]$: array of integer objects, element $C[i]$ local to process p_i in the DSM model, all elements initially zero
- $P[1 \ldots n][1 \ldots n]$: array of boolean objects, $P[i][1 \ldots n]$ local to process p_i in the DSM model, all elements initially false
- $Gate[1 \ldots n]$: array of process ID or $\bot$, $Gate[i]$ local to process p_i in DSM model, all elements initially $\bot$

Private variables:

- $incleanup$: process ID or $\bot$, uninitialized

Procedure Recover() for process p_i

```
236  if C[i] ∉ {0, 1} then
237      C[i] := 4
         // signal process in cleanup
238      incleanup := Gate[i]
239      if incleanup ≠ ⊥ then
240          P[incleanup][i] := true

         // clean up the base mutex
241      mtxA.Recover()
242      mtxA.Enter()
         // close the gate
243      for z ∈ 1 . . . n do Gate[z] := i
         // wait for processes to clear the base mutex
244      for z ∈ 1 . . . N do
245          P[i][z] := false
246          if z ≠ i ∧ C[z] ∈ {2, 3} then
247              await P[i][z]

248      mtxB.Reset()
         // open the gate
249      for z ∈ 1 . . . n do Gate[z] := ⊥
250      mtxA.Exit()
251      C[i] := 0
```

Fig. 7.3 The transformation for adding recoverability to a traditional mutex lock. Presentation continued in Fig. 7.4

Procedure Enter() for process p_i

---

252 $C[i] := 1$
     // wait at gate if needed
253 $incleanup := Gate[i]$
254 **if** $incleanup \neq \bot$ **then**
     | // signal process in cleanup
255 | $P[incleanup][i] :=$ true
     | // wait for gate to reopen
256 | **await** $Gate[i] = \bot$
257 $C[i] := 2$
258 **if** $Gate[i] \neq \bot$ **then goto** line 252
259 execute steps of $mtxB$.Enter() interleaved with reads of $Gate[i]$ until done or
         $Gate[i] \neq \bot$
260 **if** $Gate[i] \neq \bot$ **then goto** line 252

Procedure Exit() for process p_i

---

261 $C[i] := 3$
262 execute steps of $mtxB$.Exit() interleaved with reads of $Gate[i]$ until done or $Gate[i] \neq \bot$
263 $C[i] := 0$
     // signal process in cleanup
264 $incleanup := Gate[i]$
265 **if** $incleanup \neq \bot$ **then**
266 | $P[incleanup][i] :=$ true

Fig. 7.4 The transformation for adding recoverability to a traditional mutex lock. Presentation continued from Fig. 7.3

incurs additional RMRs for these processes that are in cleanup, but does not affect other processes. Assume process p_i is the one in cleanup. In the critical section of the auxiliary mutex ($mtxA$), p_i first "interrupts" the base mutex (line 243) in a manner that allows other processes to leave the base entry and exit sections in a bounded number of their own steps. It then waits using spin variables (line 247) for every other process to leave the base mutex so that it is safe to reset it. The wait goes on until other processes either arrive at a gating mechanism that prevents further access to the base mutex, or crash and recover, in both cases raising a signal by writing a spin variable. Finally, the process in cleanup resets the base mutex, opens the gate, and releases the auxiliary mutex.

The array variable $C[1 \ldots n]$ ($C[i] \in \{0, 1, 2, 3, 4\}$) records the progress of each process in a super-passage. $C[i] = 0$ implies that a process is either in NCS, or has completed the Recover procedure, or is about to enter into the NCS. $C[i] = 1$ implies that p_i has started Enter but not yet moved past the gating mechanism. $C[i] = 2$ implies that p_i has cleared the gate and is entering the base mutex $mtxB$. $C[i] = 3$ implies that p_i has completed its critical section and is about to exit the base mutex. After exiting the base lock, $C[i]$ is reset to 0. Finally, $C[i] = 4$ implies that p_i is in Recover and is about to reset the base mutex. Once p_i resets the base mutex, $C[i]$ is reset to 0.

The array $Gate[1 \ldots n]$ of spin variables is used to implement the gating mechanism such that process p_i waits on the gate using variable $Gate[i]$. The "gate" can only be closed by a process that has acquired lock $mtxA$.

In the absence of failures, a process executing the target mutex bypasses the recovery section, executes the body of the target entry section where it bypasses the gate, acquires the base mutex, $mtxB$, at line 259, completes the critical section, and finally releases the base mutex in the target exit section at line 262.

During recovery from a failure, process p_i first closes the gate at line 243 by storing its id in $Gate[j]$ for $j \in \{1, \ldots, n\}$. Any process that arrives at the gate hereafter would have to wait till the gate is opened. However, there might be processes that would have already bypassed the gate and were accessing the lock while the gate was being closed. Process p_i waits for these processes to stop accessing the lock. In order to enforce processes to stop accessing the base lock once the gate is closed, executions of $mtxB$.Enter at line 259 and $mtxB$.Exit at line 262 are modified as follows: process p_j ($j \neq i$) repeatedly checks $Gate[j]$ while accessing $mtxB$, and returns immediately to the target entry or exit section if it observes $Gate[j] \neq \bot$, meaning that the gate is closed.

The array $P[1 \ldots n][1 \ldots n]$ consists of spin/signaling variables such that p_i spins on $P[i][j]$ for $j \in \{1, \ldots, n\}$ while waiting for process p_j to leave the base lock and signal back. The base mutex is reinitialized at line 248, which is only executed after the gate is closed and every process has surrendered its access to the base lock. The structure of the algorithm guarantees that no process will access the base lock while it is being reset. The gate is later reopened at line 249 and the base mutex is then safe to be used.

Theorem 7.1 and 7.2 assert the correctness properties of the transformation.

Theorem 7.1 *The transformation described in Figs. 7.3 and 7.4 provides the 0-BR property and preserves the following correctness properties of the base mutex $mtxB$: ME, DF, SF, TE, BE and CS-continuity.*

Theorem 7.2 *If the base mutex $mtxB$ satisfies the FCFS property, then the transformation provides the 1-FCFS property.*

Theorem 7.3 asserts the complexity behavior of the transformation.

Theorem 7.3 *Suppose that the base mutex $mtxB$ has $O(f(n))$ worst-case RMR complexity per passage in the CC and DSM models for some function $f(n)$, and uses V shared variables internally. Then for any history H and any process p_i, the number of RMRs p_i incurs in the CC and DSM models in one passage is as follows:*

- Recover: *$O(n + V)$ if p_i is in cleanup after recovering with $C[i] \notin \{0, 1\}$, and $O(1)$ otherwise.*
- Enter: *$O(f(n) \times (1 + F))$ where F denotes the number of passages that begin in cleanup and interfere with p_i's passage.*
- Exit: *$O(f(n))$.*

Theorem 7.4 *The RME algorithm obtained by the transformation described in Figs. 7.3 and 7.4 is unbounded adaptive.*

7.3   Adding Boundedness

The construction presented in the previous section, when instantiated with a local-spin base mutex, has bounded RMR complexity in failure-free histories. However, the number of RMRs a process executes per passage in the worst case may be arbitrarily large as it depends on the number of crash failures in the history. This section covers an additional transformation by Golab and Ramaraju with bounded RMR complexity in the worst case, and in the absence of failures matches (up to a constant factor) the RMR complexity of the base mutex internal to the algorithm from the previous section.

The new transformation uses the construction from Sect. 7.2 as its base mutex, denoted $mtxC$, to access the target critical section. However, executions of $mtxC$.Enter must be terminated early if another process in cleanup is detected as otherwise the RMR complexity of the target entry section would be unbounded. This is accomplished by breaking out of $mtxC$.Enter, similar to an abortable mutual exclusion algorithm. Such a process may be diverted into a *bounded path* that guarantees a bounded RMR complexity. The bounded path is slower than the default path in the absence of contention, and so it is engaged only when needed.

To ensure mutual exclusion, the bounded path is protected by an auxiliary recoverable mutex, $mtxA_n$ implemented using the algorithm from Sect. 5.3. A fast path mechanism is added to $mtxA_n$ by using CAS instruction to open and close the fast path. The fast and bounded paths are synchronized using a two-process auxiliary mutex. In the absence of failures, only the fast path is accessed, which is already protected by the base mutex, and so contention is minimal. Therefore, the fast path ensures that the target mutex has the same RMR complexity asymptotically as the base mutex in the absence of failures.

Note that breaking out of $mtxC$.Enter can lead to a penalty in terms of RMRs when the process executes $mtxC$.Recover during its next passage through the target algorithm. This RMR penalty due to execution of $mtxC$.Recover can be avoided entirely if p_i breaks out of $mtxC$.Enter before line 257, where it assigns $C[i] = 2$.

The construction uses three mutexes as building blocks:

- $mtxC$ is the base mutex obtained by instantiating the construction from Sect. 7.2;
- $mtxA_n$ is the n-process auxiliary mutex that protects the critical section in the default and bounded paths, as explained earlier, implemented using the algorithm from Sect. 5.3
- $mtxA_2$ is an additional two-process auxiliary mutex used to synchronize $mtxA_n$ with the fast path, implemented using the algorithm from Sect. 5.1

In the absence of failures, a process p_i begins by executing $mtxC$.Recover, then executes $mtxC$.Enter to completion, and then attempts to "steer around" the auxiliary mutex $mtxA_n$ by executing the CAS instruction at line 275. The code after execution of line 275 in Enter where the CAS succeeds (or $G = i$ holds already prior to the CAS) and $Bounded[i]$ = false holds, and before the write operation at line 287 in Exit, is called the *fast path*.

Since the base mutex protects lines 275 to 292 in the absence of failures, there is no contention on the fast path and so p_i successfully swaps its ID into G. Next, p_i acquires $mtxA_2$ at lines 277 to 278 and enters the target critical section. After completion of the critical section, it releases $mtxA_2$ at line 286, and releases the fast path by overwriting G with $\perp$ at line 287. Thus, the fast path entails acquiring $mtxC$, successfully swapping the process id, and acquiring the auxiliary mutex $mtx\ A_2$, and then releasing all three components in the opposite order.

In the presence of failures by other processes, p_i may break out of $mtxC$.Enter at line 269. This event is recorded in $Breakout[i]$ at line 271 so that p_i remembers not to execute $mtxC$.Exit later on at line 293, as that would be unsafe given the incomplete execution of $mtxC$.Enter. On the other hand, it is safe for p_i to execute $mtxC$.Recover in its next passage. The rest of the target entry and exit sections is executed similarly to the failure-free case, except that $mtxC$ no longer protects the fast path, leading to the possibility that p_i may fail to capture the fast path at line 275 due to contention. In that case, p_i acquires $mtxA_n$ at lines 281 to 282, acquires $mtxA_2$ at lines 283 to 284, and enters the target critical section. The variable $Bounded[i]$ is used at lines 280 to 291 to record p_i's presence in the bounded path for recovery. After completion of the critical section, it releases $mtxA_2$ at line 289, and finally releases $mtxA_n$ at line 290. The code after an execution of line 275 in Enter where the CAS fails or $Bounded[i]$ = true holds, and before the write operation at line 291 in Exit, is called the *bounded path*.

If p_i itself crashes, then on recovery it executes Recover and Enter of the target lock up to line 275, as described earlier. If p_i failed inside the bounded path (i.e., $Bounded[i]$ = true), then by the test at line 275 it proceeds to line 280 where it re-enters the bounded path. In this case, p_i cannot enter the fast path at line 275 as otherwise it will acquire $mtxA_2$ with $side$ = left at lines 277 to 278 before it has a chance to recover its prior passage through $mtxA_2$ from lines 283 to 284 with $side$ = right, breaking the assumptions stated in Sect. 5.1 for accessing the two-process mutex. Otherwise, if p_i failed outside the bounded path (i.e., $Bounded[i]$ = false), then it attempts to capture the fast path, and falls back on the bounded path only if necessary. If p_i was already in the fast path when it crashed, then $G = i$ holds and the response of the CAS at line 275 indicates that it is safe for p_i to re-enter the fast path. After completing the target critical section, p_i releases the auxiliary mutexes and completes the fast path in the target exit section at lines 285 to 291.

Lemma 7.5 *For any history H,*

1. *at most one process is in the fast path at a time*
2. *at most one process in the bounded path is able to acquire mtx_{A_n} at a time*

Theorem 7.6 asserts the correctness properties of the transformation.

Shared variables:

- $mtxA_2$: two-process recoverable mutex implemented using algorithm in fig. 5.1 from section 5.1, and augmented with BCSR using fig. 5.2 from section 5.2
- $mtxA_n$: n-process recoverable mutex implemented using algorithm in fig. 5.3 from section 5.3, and augmented with BCSR using fig. 5.2 from section 5.2
- $mtxC$: recoverable base mutex implemented using algorithm from section 7.2, and augmented with BCSR using fig. 5.2 from section 5.2
- G: holds process ID or $\bot$, initially $\bot$, determines state of fast path to $mtxA_N$

Private variables (stored in non-volatile memory):

- $Bounded[1\ldots n]$: array of boolean, all elements initially false
- $Breakout[1\ldots n]$: array of boolean, element i private to process p_i, all elements initially false

---

Procedure Recover() for process p_i

---

267   $mtxC$.Recover()

Procedure Enter() for process p_i

---

268 **if** $\neg Breakout[i]$ **then**
269 | execute steps of $mtxC$.Enter(), break out immediately after the second completion of lines 252 to 256 of algorithm from section 7.2
270 **if** broke out of $mtxC$.Enter() at line 269 **then**
271 | $Breakout[i] :=$ true
272 **else**
273 | $Breakout[i] :=$ false

    // try to capture the fast path using CAS
274 **if** $\neg Bounded[i]$ **then**
275 | CAS($\&G$, $\bot$, i)
276 | **if** $G = i$ **then**
         // fast path of default path
277 | | $mtxA_2$.Recover(left)
278 | | $mtxA_2$.Enter(left)
279 | | **return**

    // bounded path
280 $Bounded[i] :=$ true
281 $mtxA_n$.Recover()
282 $mtxA_n$.Enter()
283 $mtxA_2$.Recover(right)
284 $mtxA_2$.Enter(right)

Fig. 7.5 The transformation to guarantee bounded RMR complexity. Presentation continued in Fig. 7.6

Procedure Exit() for process p_i

---

285 **if** $G = i$ **then**
286 | $mtxA_2$.Exit(left)
        | // release the fast path
287 | $G := \bot$
288 **else**
289 | $mtxA_2$.Exit(right)
290 | $mtxA_N$.Exit()
291 | $Bounded[i] :=$ false
292 **if** $\neg Breakout[i]$ **then**
293 | $mtxC$.Exit()
294 **else**
295 | $Breakout[i] :=$ false

Fig. 7.6 The transformation to guarantee bounded RMR complexity. Presentation continued from Fig. 7.5

Theorem 7.6 *The transformation preserves the following correctness properties of the base mutex mutxC: ME, DF, SF, JJ-SF, BCSR, TE, BE, 0-BR and CS continuity.*

Theorem 7.7 asserts the complexity behavior of the transformation.

Theorem 7.7 *Suppose that the base mutex internal to $mtxC$ has $O(f(n))$ worst-case RMR complexity per passage in the CC and DSM models for some function $f(n)$, and uses V shared variables internally. Then, the number of RMRs a process p_i incurs in one passage through this transformation in the CC and DSM models is $O(n + V + f(n))$ if p_i's passage is 2-failure-concurrent, and $O(f(n))$ otherwise.*

Theorem 7.8 *The RME algorithm obtained by the transformation described in Figs. 7.5 and 7.6 is bounded semi-adaptive.*

7.4   Weak Recoverability

To design a well-bounded super-adaptive RME algorithm, Dhoked and Mittal use a solution to the *weaker* variant of the RME problem as a *building block* in which a failure may cause the ME property to be violated albeit only temporarily and in a controlled manner. This variant is referred to as the *weakly recoverable mutual exclusion problem*.

To formally define how long a violation of the ME property may last, the notion of consequence interval of a failure is used. Roughly speaking, the notion of consequence interval of a failure captures the maximum duration for which the impact of the failure may

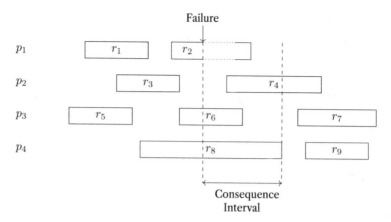

Fig. 7.7 Illustration of the consequence interval of a failure

be felt in the system. It is related to, but different from, the notion of k-failure-concurrent passage. These differences are discussed in Sect. 7.7.

Definition 7.1 *(consequence interval)* The *consequence interval* of a failure in a history H is defined as the interval in time that starts from the onset of the failure and extends to the point in time when either every super-passage that started before this failure occurred in H has completed or the last step in H is performed, whichever happens earlier.

Figure 7.7 demonstrates the consequence interval of a failure. Based on this notion, the weakly RME is defined.

Definition 7.2 *(Weakly Recoverable Mutual Exclusion)* An algorithm is a *weakly recoverable mutual exclusion algorithm* if, in addition to starvation freedom, it satisfies the following property: for any history H, if two or more processes are in their critical sections simultaneously at some point in H, then that point overlaps with the consequence interval of some failure.

Roughly speaking, a weakly RME algorithm satisfies the ME property as long as no failure has occurred in the "recent" past. The bounded exit, bounded recovery and bounded critical section reentry properties are accordingly applicable to weakly RME problem as well.

The notion of consequence interval is also used to define a new fairness property, CI-FCFS.

Definition 7.3 *(CI-FCFS)* A history H is said to satisfy CI-FCFS if it satisfies the following property. Consider a pair of passages ℓ_i and ℓ_j in H belonging to processes p_i and p_j, respectively, such that (a) p_i completes its doorway in ℓ_i before p_j begins its doorway in ℓ_j, and (b) ℓ_i does not overlap with the consequence interval of any failure. If p_j has started the CS in ℓ_j, then p_i has started the Exit section in ℓ_i.

The next section demonstrates an optimal weakly recoverable mutual exclusion algorithm using existing hardware instructions whose worst-case RMR complexity is only $O(1)$ under both CC and DSM models.

7.5   An Optimal Weakly Recoverable Lock

This section covers a weakly recoverable lock whose RMR complexity is $O(1)$ per passage irrespective of the number of failures under both CC and DSM models. This lock is based on the well-known MCS queue-based (non-recoverable) lock [4]. For this algorithm, the MCS lock is augmented to satisfy the bounded-exit property, and to make it weakly recoverable.

Pseudocode of the algorithm that implements a weakly recoverable lock is given in Figs. 7.8, 7.9 and 7.10.

The lock works as follows. The algorithm maintains a first-come-first-served (FCFS) queue of outstanding requests using a linked-list of their associated nodes. A node contains two fields: (a) *next*, a reference to its successor node in the queue (if any), and (b) *locked*, a boolean variable used by a process to spin while waiting for its turn to enter its critical section. The queue itself is represented using a shared variable *tail* that contains a reference to the last node in the queue if non-empty and **null** otherwise.

Additionally each process p_i maintains three more variables,

- *pred*[i]: address of the predecessor node, if any, of process p_i after its node has been appended to the queue
- *mine*[i]: address of the queue node associated with process p_i's most recent request
- *state*[i]: process p_i's current state. The state of a process has five possible values, namely FREE, INITIALIZING, TRYING, INCS and LEAVING. The state information is used for recovery from failures as explained later.

To acquire the lock, a process first initializes its queue node by setting its *next* and *locked* fields to **null** (line 306) and true (line 307), respectively. It then appends the node to the queue by performing an FAS instruction on *tail* at line 312. Note that this instruction returns the contents of *tail* just before it is modified. If the return value is **null**, then it indicates that the lock is free and the process has successfully acquired the lock. If not, then it indicates that the lock is not free and the return value is the reference to the predecessor of the process' own node in the queue. In that case, it notifies the owner of the predecessor node of its presence. To that end, it stores the reference to its own node in the *next* field of the predecessor node using a CAS instruction (line 316), thereby creating a forward link

Shared variables:

- *tail*: pointer to QNode, initially **null**
- *state*[1...*n*]: array of integer objects, all elements initially FREE
- *mine*[1...*n*]: array of pointers to QNode, all elements initially **null**
- *pred*[1...*n*]: array of pointers to QNode, all elements initially **null**

Private variables:

- *result*: pointer to QNode, uinitialized

Procedure Recover() for process p_i

```
296  if state[i] = TRYING then
297        if pred[i] = mine[i] then
                 /* Failed while performing FAS instruction; abort the attempt. If FAS
                    was successful, the two references are guaranteed to be different.
                 */
298            Cleanup()                                      // execute cleanup method
299  else if state[i] = LEAVING then
300        Cleanup()                                          // finish executing cleanup method
301  if state[i] = FREE then                                  // initialize lock
302        state[i] := INITIALIZING                           // advance the state
```

Fig. 7.8 The algorithm for a weakly recoverable lock. Presentation continued in Figs. 7.9 and 7.10

between the two nodes. If this link is created successfully, the process starts spinning on the *locked* field (line 318) of its own node waiting for it to be reset to false by the owner of the predecessor node as part of releasing the lock. Otherwise, the predecessor has completed its Enter and proceeds to enter its CS.

The lock is released with the help of the procedure Cleanup. To release the lock, a process first tries to reset the *tail* variable to **null** using a CAS instruction at line 323. Then, it attempts to store a special value (e.g., reference to its own node) in the *next* field of its own node using a CAS instruction at line 324. If this CAS is unsuccessful, then the process can conclude that a forward link has already been created. It then follows this link and resets the *locked* field of its successor node (line 326). On the other hand, if the CAS instruction is successful, then the exiting process doesn't need to do anything and can simply return. This CAS instruction (line 313) is designed to synchronize with the CAS instruction used to create the forward link (line 316), such that only one of the two instructions can be successful, thereby ensuring that the *next* field can only be modified once. This mechanism allows a leaving process to notify the process next in line that the lock is now free, in case a forward link is not established already.

In case of failures, the algorithm works as follows. Recall that a process uses the FAS to append its own node to the queue and also obtain the address of its predecessor node. If a process fails immediately after executing this FAS instruction (line 312) but before it is able to store its return value to the shared memory (line 313), there is no easy way to recover this address (of the predecessor) based on the current knowledge of the failed process. The

---

Procedure Enter() for process p_i

---

303 **if** $state[i]$ = INITIALIZING **then**
304   **if** $mine[i] = $ **null then**
305          $\lfloor$ $mine[i] := $ a new QNode object

         /* initialize fields of my own node   */
306      $mine[i].next := $ **null**
307      $mine[i].locked := $ true
         /* To determine success of FAS   */
308      $pred[i] := mine[i]$
309      $state[i] := $ TRYING   // advance the state
310 **if** $state[i]$ = TRYING **then**
311   **if** $pred[i]$ = $mine[i]$ **then**
             /* append my own node to the queue   */
312          $result := $ FAS ($tail, mine[i]$)
             /* persist the result of FAS   */
313      $\lfloor$ $pred[i] := result$

314 **if** $state[i]$ = TRYING **then**
315   **if** $pred[i] \neq $ **null then**
             /* have a predecessor; create the link   */
316   CAS (&$pred[i].next$, **null**, $mine[i]$)
317   **if** $pred[i].next$ = $mine[i]$ **then**
                 /* wait for the predecessor to complete   */
318              $\lfloor$ **await** $\neg mine[i].locked$ // spin
319      $state[i] := $ INCS   // advance the state

Procedure Exit() for process p_i

---

320 Cleanup()

Fig. 7.9 The algorithm for a weakly recoverable lock. Presentation continued from Fig. 7.9 and continued in Fig. 7.10

Procedure Cleanup() for process p_i

---

321 $state[i] := $ LEAVING   // advance the state
322 **if** $mine[i] \neq $ **null then**
         /* remove my node from the queue if it has no successor   */
323   CAS (&$tail, mine[i]$, **null**)
         /* may have a successor; signal it to enter CS   */
324   CAS (&$mine[i].next$, **null**, $mine[i]$)
325   **if** $mine[i].next \neq mine[i]$ **then**
             /* link already created; tell the successor to stop spinning */
326          $\lfloor$ $mine[i].next.locked := $ false
327      $pred[i] := $ **null**
328      $mine[i] := $ **null**
329 $state[i] := $ FREE   // advance the state

Fig. 7.10 The algorithm for a weakly recoverable lock. Presentation continued from Figs. 7.8 and 7.9

queue continues to grow beyond this node, but it would be disconnected from the previous part of the queue, thereby creating one more sub-queue. For an example, please refer to Fig. 7.11.

If a process detects that it may have failed while executing the FAS instruction at line 296, it "relinquishes" its current node, informs its successor (if any) that the lock is now "free" using the Cleanup procedure described earlier and retries acquiring the lock using a new node. This potentially creates multiple queues (or sub-queues) which may allow multiple processes to execute their critical sections concurrently, thereby violating the ME property. This makes the lock a weakly recoverable lock.

If the process fails anywhere else, the recovery uses the $state[i]$ information to detect where the process failed and simply reruns that method since the rest of the algorithm is idempotent, in the sense that multiple runs of the same procedure would yield the same result as a single run of that procedure.

Note that each failure splits the queue into at most two subqueues. Additionally, at most one process from each subqueue can be in the CS. Thus, the following theorems follow.

Lemma 7.9 *Given a history H, and a non-negative integer F, if F + 1 processes are in their critical sections simultaneously at some point, then there exists at least F failures whose consequence interval overlaps with that point.*

Theorem 7.10 *The algorithm described in Figs. 7.8, 7.9 and 7.10 satisfies the JJ-SF, BCSR, BE, BR and CS continuity properties.*

Theorem 7.11 *The algorithm described in Figs. 7.8, 7.9 and 7.10 has RMR complexity of O (1) for each of* Recover, Enter *and* Exit *sections in the CC and DSM models.*

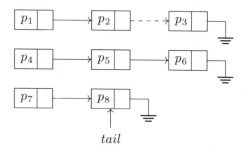

tail

Fig. 7.11 Processes $p_1, \ldots, p_8$ successfully append their nodes to the tail of the queue using an FAS instruction. Processes p_4 and p_7 failed to store the outcome of the FAS instruction to persistent memory, and are thus unable to set the next field of the nodes of p_3 and p_6. Process p_3 has captured the address of the node of p_2 and is about to set the corresponding next field on the node of p_2. Effectively, three sub-queues are created due to failures of p_4 and p_7

7.6   A Strongly Recoverable Well-Bounded Super-Adaptive Lock

This section describes a framework that uses other types of recoverable locks as building blocks to construct a lock that is not only strongly recoverable but also well-bounded super-adaptive under both CC and DSM models. This (well-bounded super-adaptive) lock is illustrated in two steps. First, a basic framework that transforms a bounded non-adaptive strongly recoverable lock to a bounded semi-adaptive strongly recoverable lock. Then, this framework is extended to make the lock super-adaptive while ensuring that it stays strongly recoverable and bounded. Finally, instantiating the framework with an appropriate well-bounded non-adaptive lock yields the desirable well-bounded super-adaptive lock.

7.6.1   A Well-Bounded Semi-Adaptive RME Algorithm

This framework is based on the one used by Golab and Ramaraju as shown in Sect. 7.3 in order to construct a strongly recoverable lock that is semi-adaptive. In this framework, in the presence of failures (even a single failure), the RMR complexity of the target lock is dominated by the overhead of aborting the attempt to acquire the base lock and then resetting the base lock, thereby making the lock semi-adaptive.

This framework consists of four different components as building blocks.

- *Filter lock:* The weakly recoverable lock described above that provides mutual exclusion in the absence of failures
- *Splitter:* Used to split processes into *fast* or *slow* paths. If multiple processes navigate the splitter concurrently (which would happen only if a failure has occurred), at most one of them is allowed to take the fast path and the rest are diverted to the slow path. It is implemented using an atomic integer and a CAS instruction.
- *Arbitrator lock:* A *2-ported* strongly recoverable lock from Sect. 5.1. Each port corresponds to a side. The two sides are left and right. At any time, at most one process is allowed to attempt to acquire the lock from any side. However, *any two* of the n processes can compete to acquire the lock.
- *Core lock:* a (presumably non-adaptive) strongly recoverable lock that assures mutual exclusion among processes taking the slow path

In order to acquire the target lock, a process proceeds as follows. It first waits to acquire the filter lock. Once granted, it navigates through the splitter, trying to enter the fast path. If successful, it then attempts to acquire the arbitrator lock from the left side. If multiple processes acquire the filter lock simultaneously, due to failures, it results in contention at the splitter, where all but one processes are diverted to the slow path. If a process is forced to take the slow path, it attempts to acquire the core lock. Once granted, it then waits to acquire the arbitrator lock from the right side. Finally, once the process has successfully acquired the arbitrator lock, it is deemed to have acquired the target lock as well, and is now in the CS of the target lock. A pictorial representation of the execution flow is depicted in Fig. 7.12. Pseudocode of the transformation is given in Fig. 7.13.

In the absence of failures, every process takes the fast path. Note that, generally, at most one process can take the fast path at a time and at most one process can hold the core lock at a time. Any process that takes the fast path will always attempt to acquire the arbitrator lock from the LEFT side. Any process that takes the slow path and acquires the core lock will always attempt to acquire the arbitrator lock from the RIGHT side. Since the core lock is strongly recoverable, at most one process will try to acquire the arbitrator lock from each side at a time.

In order to release the target lock, a process simply releases its component locks in the reverse order in which it acquired them: the arbitrator lock, followed by the core lock (in case the process took the slow path), followed by the splitter and finally the filter lock.

We refer to the algorithm described in this subsection as SA- LOCK. The doorway of SA- LOCK is given by the doorway of its filter lock. When convenient, we use $\mathcal{F}$ and $\mathcal{C}$ to refer to the filter and core locks, respectively, of SA- LOCK.

Theorem 7.12 asserts the correctness properties of SA- LOCK.

Theorem 7.12 SA- LOCK *provides the CI-FCFS property and preserves the following correctness properties of the core lock: ME, JJ-SF, BCSR, TE, BE, BR, k-BR and CS continuity properties.*

Theorem 7.13 asserts the complexity behaviour of SA- LOCK.

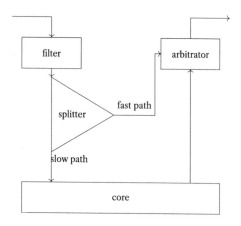

Fig. 7.12 A pictorial representation of the framework

Shared variables:

- $\mathcal{F}$: n-process weakly recoverable lock
- *owner*: integer object to implement splitter - used to store the identifier of the process currently occupying the fast path, initially 0
- $\mathcal{C}$: core lock, n-process strongly recoverable lock
- $\mathcal{A}$: arbitrator lock, 2-ported n-process strongly recoverable lock
- *type*[1...*n*]: array of boolean to store the path taken by the process, each element in the set {FAST, SLOW}, all elements initially FAST

Definitions:

$$side(type) = \begin{cases} \text{Left} & \text{if } type = \text{FAST} \\ \text{Right} & \text{otherwise} \end{cases}$$

Procedure `Recover()` for process p_i

```
/* no special recovery actions are needed                              */
```

Procedure `Enter()` for process p_i

330	$\mathcal{F}$.Recover()	// recover the filter lock
331	$\mathcal{F}$.Enter()	// acquire the filter lock
332	**if** $type[i] \neq$ SLOW **then**	// not yet on the slow path
333	$\quad$ CAS(&*owner*, 0, *i*)	// attempt to take the fast path
334	**if** *owner* $\neq i$ **then**	// unable to take the fast path
335	$\quad$ *type*[*i*] := SLOW	// committed to take the slow path
336	$\quad$ $\mathcal{C}$.Recover()	// recover the core lock
337	$\quad$ $\mathcal{C}$.Enter()	// acquire the core lock
338	$\mathcal{A}$.Recover($side(type[i])$)	// recover arbitrator lock
339	$\mathcal{A}$.Enter($side(type[i])$)	// acquire the arbitrator lock

---

Procedure `Exit()` for process p_i

340	$\mathcal{A}$.Exit($side(type[i])$)	// release the arbitrator lock
341	**if** $type[i]$ = SLOW **then**	// took the slow path
342	$\quad$ $\mathcal{C}$.Exit()	// release the core lock
343	**else**	// took the fast path
344	$\quad$ *owner* := 0	// the fast path is now empty
345	*type*[*i*] := FAST	// reset the path type to default
346	$\mathcal{F}$.Exit()	// release the filter lock

Fig. 7.13 The transformation to convert a bounded non-adaptive strongly recoverable lock to a bounded semi-adaptive strongly recoverable lock

Theorem 7.13 (SA- LOCK is bounded semi-adaptive) SA- LOCK *has RMR complexity of* $O\,(1)$ *per passage if its associated super-passage does not overlap with the consequence internal of any failure and* $O\,(R(n))$ *otherwise in the CC and DSM models, where* $R(n)$ *denotes the worst-case RMR complexity of the core lock for n processes.*

Theorem 7.14 builds from Lemma 7.9 and establishes the potential adaptive nature of the SA- LOCK that is leveraged to construct a well-bounded super-adaptive RME algorithm in Sect. 7.6.2.

Lemma 7.14 *Given a history H, and a non-negative integer F, if F processes are competing for the core lock at some point, then there exists at least F failures whose consequence interval overlaps with that point.*

7.6.2   A Well-Bounded Super-Adaptive RME Algorithm

Dhoked and Mittal capitalize on the *gap* between the worst-case RMR complexity of implementing a weakly recoverable lock and that of implementing a strongly recoverable lock to achieve a well-bounded super-adaptive RME algorithm.

The main idea is to *recursively* transform the core lock using instances of the semi-adaptive lock repeatedly up to a height m that is equal to the worst-case RMR complexity of another strongly recoverable lock under arbitrary number of failures. The strongly recoverable lock now becomes the base case of the recursion. For ease of exposition, we refer to the core lock in the base case as the *base lock*.

Let NA- LOCK be a bounded (presumably non-adaptive but does not have to be) strongly recoverable lock, whose worst-case RMR complexity is $O(R(n))$ for n processes. Let SA- LOCK denote an instance of the semi-adaptive lock described in Sect. 7.6.1. And, finally, let BA- LOCK denote the bounded super-adaptive lock that we wish to construct (NA- LOCK is the base lock and BA- LOCK is the target lock).

The idea is to create $m = R(n)$ levels of SA- LOCK such that the core lock component of the SA- LOCK at each level is built using another instance of SA- LOCK for up to $m - 1$ levels and using an instance of NA- LOCK at the base level (level m). Let SA- LOCK$[i]$ denote the instance of SA- LOCK at level i. Formally,

$$\text{BA- LOCK} = \text{SA- LOCK}[1]$$
$$\text{SA- LOCK}[i].\text{core} = \text{SA- LOCK}[i + 1] \quad \forall i \in \{1, 2, \ldots, m - 1\}$$
$$\text{SA- LOCK}[m].\text{core} = \text{NA- LOCK}$$

A pictorial representation of the execution flow of the recursive framework is depicted in Fig. 7.14.

In order to acquire the target lock, a process starts at the first level and waits to acquire the filter lock at level 1. It stays on track to acquire the fast path until a failure occurs with respect to the filter lock at the first level as a result of which multiple processes may be granted access to the (filter) lock simultaneously. All of these processes then compete to

Fig. 7.14 A pictorial
representation of the recursive
framework

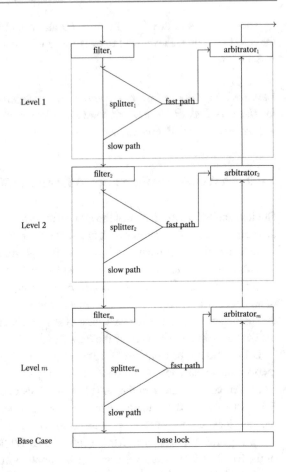

enter the fast path by navigating through the splitter. The splitter allows only one process to take the fast path at a time, and the rest are diverted to take the slow path.

All processes in the slow path at the first level then move to the second level. If no further failure occurs, then no process takes the slow path at the second level and all processes leave this level one-by-one using the fast path with respect to this level. Thus, only $O(1)$ RMR complexity is *added* to the passages of all the affected processes until the impact of the first failure has subsided. However, if one or more processes take the slow path at the second level, then we can infer that a *new* failure must have occurred with respect to the filter lock at the second level. All these processes at the slow path of the second level then move to the third level, and so on and so forth. At each level, a process on the slow path, upon either acquiring the base lock or returning from the adjacent higher level (whichever case applies), it next waits to acquire the level-specific arbitrator lock. Once granted, it either returns to the adjacent lower level or, if at the initial level, is deemed to have successfully acquired the target lock. As before, in order to release the target lock, a process releases its components locks in the reverse order in which it acquired them.

Note that in this algorithm, at least k failures are required at any level to force k processes to be "escalated" to the next level (from Theorem 7.14). Each level except for the last one would add only $O(1)$ RMR complexity to the passages of these process, thus making the target lock adaptive under limited failures. There is no further "escalation" of processes at the base level and a bounded (possibly non-adaptive) strongly recoverable lock is used to manage all these processes at that point, thus bounding its RMR complexity under an arbitrarily large number of failures as well.

Theorem 7.15 asserts the correctness properties of SA-LOCK.

Theorem 7.15 BA-LOCK *satisfies the ME, JJ-SF, BCSR, BE, BR and CI-FCFS properties.*

Lemma 7.16 *Suppose a process p advances to level x at some time t during its super-passage, where $1 \leq x \leq m$. Then, there exist at least $\frac{x(x-1)}{2}$ failures that occurred at or before time t whose consequence interval overlaps with the super-passage of the process p.*

Theorem 7.17 asserts the complexity behaviour of SA-LOCK.

Theorem 7.17 (BA-LOCK is bounded super-adaptive) BA-LOCK *has RMR complexity of $O\left(\min\{\sqrt{F+1}, R(n)\}\right)$ per passage in the CC and DSM models, where F denotes the number of failures whose consequence interval overlaps with the super-passage associated with that passage and R(n) denotes the RMR complexity of the base NA-LOCK for n processes.*

Corollary 7.18 (a well-bounded super-adaptive lock) *Assume that we use a sub-logarithmic algorithm from Chap. 6 to implement the NA-LOCK. Then the RMR complexity of any passage in a super-passage of the BA-LOCK is given by $O\left(\min\{\sqrt{F+1}, \log n/\log\log n\}\right)$, where F denotes the number of failures whose consequence interval overlaps with the super-passage.*

7.7   Relation Between with CI-FCFS and k-FCFS

Section 4.1 defines the concept of a k-failure-concurrent passage, where $k \geq 0$, to capture how a failure may impact the RMR complexity of a passage. On the other hand, this chapter uses the concept of consequence interval of a failure, defined in Sect. 7.4. Given a history, a passage is said to be *CI-concurrent* if its super-passage overlaps with the consequence interval of one or more failures. A natural question is to ask how the concept of k-failure-concurrent is related to CI-concurrent. We show that CI-concurrent is "stronger" than 2-failure-concurrent and "incomparable" with 1-failure-concurrent.

Theorem 7.1 *Given a passage r in a finite history H, if r is CI-concurrent, then it is also 2-failure-concurrent.*

Proof For convenience, given a passage r, we use $\Pi(r)$ to denote the super-passage associated with r.

Consider a passage r_i of process p_i that is CI-concurrent. By definition, $\Pi(r_i)$ overlaps with the consequence interval of a failure, say f.

Let f be the failure of process p_j while executing the passage r_j.

We use $\mathbb{P}(f)$ to denote the set of processes that have a super-passage in progress at the time when f occurred. Note that $\mathbb{P}(f) \neq \emptyset$ because $p_j \in \mathbb{P}(f)$. Let $p_k \in \mathbb{P}(f)$ denote the process whose super-passage, which was in progress when f occurred, extends for the *longest* time in H. Finally, let r_k denote the *most recent* passage of p_k that started before f occurred.

Note that $\Pi(r_k)$ overlaps with $\Pi(r_j)$. This in turn implies that r_k is 1-failure-concurrent because r_j is 0-failure-concurrent. Also, note that $\Pi(r_k)$ contains the consequence interval of f and hence overlaps with $\Pi(r_i)$. This implies that r_i is 2-failure-concurrent. $\qquad\square$

The following examples to demonstrate that 1-failure-concurrent and CI-concurrent are incomparable properties, i.e., neither implies the other.

Theorem 7.2 *There exists a history H and a passage r in H such that r is 1-failure-concurrent but not CI-concurrent.*

Proof Consider the history shown in Fig. 7.15. The passage r_3 in the history is 1-failure-concurrent but not CI-concurrent.

Fig. 7.15 Passages r_1 and r_2 are the only 0-failure-concurrent passages in the history. Passage r_3 overlaps r_1 and thus is 1-failure-concurrent, but is not CI-concurrent

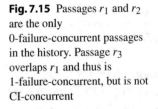
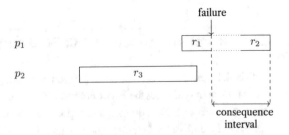

Fig. 7.16 Passages r_1 and r_2 are the only 0-failure-concurrent passages in the history. Passage r_4 overlaps with the consequence interval of the failure and is 2-failure-concurrent, but not 1-failure-concurrent

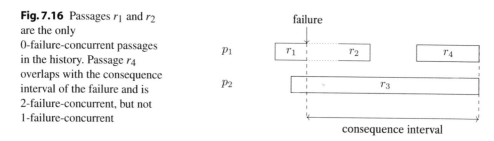

Theorem 7.3 *There exists a history H and a passage r in H such that r is CI-concurrent but not 1-failure-concurrent.*

Proof Consider the history shown in Fig. 7.16. The passage r_4 in the history is CI-concurrent but not 1-failure-concurrent. □

Finally, we show that CI-concurrent is a strictly stronger property than 2-failure-concurrent, i.e., the latter does not imply the former.

Corollary 7.4 *There exists a history H and a passage r in H such that r is 2-failure-concurrent but not CI-concurrent.*

Section 4.3 uses the notion of k-failure concurrency to define a notion of fairness especially suited to RME algorithms, referred to as k-FCFS. Intuitively, k-FCFS guarantees FCFS among two passages provided none of them is k-failure-concurrent.

The results above also establish the following relationship among different fairness properties (a) CI-FCFS implies 2-FCFS, and (b) CI-FCFS and 1-FCFS are incomparable.

References

1. Sahil Dhoked and Neeraj Mittal. An adaptive approach to recoverable mutual exclusion. In *Proc. of the 39th ACM Symposium on Principles of Distributed Computing (PODC)*, PODC '20, pages 1–10, New York, NY, USA, 2020. Association for Computing Machinery.
2. Wojciech Golab and Aditya Ramaraju. Recoverable mutual exclusion. In *Proc. of the 35th ACM Symposium on Principles of Distributed Computing (PODC)*, pages 65–74, 2016.
3. Wojciech Golab and Aditya Ramaraju. Recoverable mutual exclusion. *Distributed Computing (DC)*, 32(6):535–564, 2019.
4. John M. Mellor-Crummey and Michael L. Scott. Algorithms for scalable synchronization on shared-memory multiprocessors. *ACM Transactions on Computer Systems (TOCS)*, 9(1):21–65, 1991.

Constant Amortized Complexity Algorithm

8

In [1], Chan and Woelfel describe an RME algorithm that has *constant* amortized RMR complexity. In addition to read and write instructions, it uses the *fetch-and-increment* (FAI) as well as *compare-and-swap* (CAS) read-modify-write instructions, which are commonly available on most modern processors including Intel 64 and AMD64. As explained in Sect. 4.7, this instruction takes a memory location as input, increments the contents of the location by one, and also returns the previous contents of the location.

8.1 The Main Idea

The algorithm is based on Anderson's array-based mutual exclusion algorithm that has $O(1)$ worst-case RMR complexity under both CC and DSM models. It uses the notion of *tickets*. A process, upon generating a request, first obtains a ticket number, then proceeds to claim the ticket associated with the number and, if successful, waits for its turn to be serviced. Requests are serviced in the order of their ticket numbers. A ticket is said to have been used once the associated request (to which the ticket was issued) has been fulfilled.

Each ticket is mapped to a unique slot in a shared (infinite) *ticket array* using its number. The ticket array is used to store information about the current status of each ticket. The status of a ticket may be *unclaimed*, *claimed* or *canceled*. Initially, all tickets are unclaimed. Once a ticket has been claimed, its associated slot stores information about the ticket holder (*i.e.*, the process to which the ticket belongs).

A shared *ticket counter* is used to issue ticket numbers to processes in a sequential manner. A process obtains a ticket number by incrementing the shared counter using the FAI instruction. After obtaining the ticket number, a process claims the ticket by updating the associated slot in the ticket array. A token, which is also a shared counter, is used to keep

© The Author(s), under exclusive license to Springer Nature Switzerland AG 2023
S. Dhoked et al., *Recoverable Mutual Exclusion*, Synthesis Lectures on
Distributed Computing Theory, https://doi.org/10.1007/978-3-031-20002-1_8

track of the ticket number that is currently being serviced. Mutual exclusion is achieved by requiring that only the holder of the ticket currently pointed to by the token can execute its critical section. Starvation freedom is achieved by advancing the token once the ticket corresponding to the current token value has been used or canceled and there is a pending request in the system. The system contains a pending request if the two counters (*ticket* and *token*) are not equal.

Note that the return value of the FAI instruction is stored in a CPU register, which is not persistent and needs to be copied explicitly to shared memory for persistence. As a result, if a process crashes while executing the FAI instruction, then, upon recovery, it is not possible for the process to tell if it crashed before or after executing the FAI instruction. In the latter case, the ticket number generated by the process before its failure is considered to be lost (forever).

This loss of a ticket number due to a process failure can block the token from advancing further, unless a mechanism is used to cancel the associated ticket (whose token number has been lost) so that the token can be advanced. To enable that, while trying to advance the token, if a process detects that a failure has occurred *and* the ticket associated with the current token value has not been claimed yet, then it proceeds to cancel the ticket by changing its status in the ticket array. However, this *may* cause even a valid ticket (whose ticket number has not been lost) to be canceled if its holder is too slow to claim the ticket.

While trying to claim a ticket, if a process finds that the ticket has already been canceled, it simply retries. To that end, it obtains a new ticket number and retries the steps outlined earlier.

Chan and Woelfel's algorithm guarantees a ticket associated with a lost ticket number is eventually canceled. However, to minimize spurious cancellations, it also ensures that the number of valid tickets that are canceled is upper-bounded by a linear function of the number of failures that have occurred. In particular, it guarantees that no ticket is canceled if no process fails.

8.1.1   Advancing the Token

The procedure used to advance the token is both concurrency safe and failure safe; multiple processes can execute it concurrently or a process can execute it multiple times without any adverse effect. The token value itself is advanced using a CAS instruction. A process executes this procedure upon arrival (after claiming its ticket), departure (after leaving its critical section) or recovery (after a failure). The procedure is invoked repeatedly as long as the required conditions for advancing the token are met. To obtain constant amortized RMR complexity, if multiple processes invoke the procedure and try to advance the token using a CAS instruction, only the process whose CAS instruction succeeds is allowed to invoke the procedure again.

8.2   Formal Description

A more formal description of the algorithm is given in Figs. 8.1 and 8.2. The algorithm has been modified from its original version to adhere to the Golab and Ramaraju's execution model in two ways. In every passage, a process executes `Recover`, `Enter` and `Exit` procedures in order (with critical section in between `Enter` and `Exit`). Second, there are no "jump" statements from one procedure to another, which are not allowed in a high-level programming language such as C/C++.

Figure 8.1 shows the shared variables used by the algorithm as well as pseudocode for the `Recover`, `Enter` and `Exit` procedures. Figure 8.2 shows the pseudocode for the two helper procedures `WaitingRoom` and `MoveToken`.

The algorithm uses three shared counters, namely the ticket counter C, shadow counter D and token counter *Token*. The first counter C is used to generate ticket numbers. The second counter D is used to keep track of the number of times the processes have either claimed their tickets or crashed. The third counter *Token* is used to keep track of the number of tickets that have been either used or canceled. In the absence of failures, the shadow counter's value is upper-bounded by that of the ticket counter. However, in the presence of failures, its value may exceed that of the ticket counter and, only then, one or more tickets may be canceled. The infinite array A is used to store the status of each ticket. The algorithm also uses three bounded arrays of size n with one entry for each process, namely *Status*, *W* and *Exit*. The first array *Status* is used to indicate if a process is executing its super-passage. The second array *W* is used to record the current ticket number of a process, if it is executing

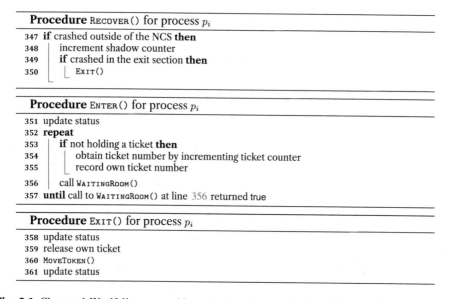

Procedure RECOVER() for process p_i

347	**if** crashed outside of the NCS **then**
348	increment shadow counter
349	**if** crashed in the exit section **then**
350	EXIT()

Procedure ENTER() for process p_i

351	update status
352	**repeat**
353	**if** not holding a ticket **then**
354	obtain ticket number by incrementing ticket counter
355	record own ticket number
356	call WAITINGROOM()
357	**until** call to WAITINGROOM() at line 356 returned true

Procedure EXIT() for process p_i

358	update status
359	release own ticket
360	MOVETOKEN()
361	update status

Fig. 8.1 Chan and Woelfel's recoverable mutual exclusion algorithm. Presentation continued in Fig. 8.2

Procedure WaitingRoom() for process p_i

---

362 **if** own ticket not yet claimed **then**
363   try to claim own ticket in ticket array
364   **if** ticket was cancelled **then**
365   release own ticket
366   **return** false
367   increment shadow counter
368   MoveToken()
369 **await** own ticket has been serviced
370 **return** true

Procedure MoveToken() for process p_i

---

371 read current token
372 read shadow counter and ticket counter
373 **if** token is behind the ticket counter **then**
374   read next ticket after token in ticket array
375   **if** next ticket already claimed **then**
376   try to service next ticket
377   **if** next ticket serviced and belongs to p_i **then**
378   try to advance token
379   **if** successful **then**
380   MoveToken()
381   **else**
382   **if** ticket counter not ahead of shadow counter **then**
383   try to cancel next ticket after token in ticket array
384   **if** next ticket canceled **then**
385   try to advance token
386   **if** successful **then**
387   MoveToken()

Fig. 8.2 Chan and Woelfel's recoverable mutual exclusion algorithm. Presentation continued from Fig. 8.1

a super-passage. Finally, the third array *Exit* is used to indicate if a process is executing its Exit section.

In the Recover procedure, the process checks whether its previous passage ended up in failure. If so, it increments the shadow counter (line 348) and completes the Exit procedure, if applicable (line 350).

In the Enter procedure, a process generates a ticket number (line 354), writes it to shared memory to make it persistent (line 355), and enters the waiting room by invoking the WaitingRoom procedure (line 357). In the waiting room, the process tries to claim the ticket (line 363) and, if successful, increments the shadow counter (line 367), attempts to advance the token (line 368) and then waits for its turn to be serviced (lines 369). However, if the ticket is canceled before the process is able to claim it (line 364), the process releases the ticket (line 365) and leaves the waiting room (line 366).

In the Exit procedure, the process releases the ticket number (line 359) and attempts to advance the token by invoking the MoveToken procedure (line 360).

In the `MoveToken` procedure, the process reads the values of the three counters into its private memory (lines 371 to 372). If the token counter value lags the ticket counter value (line 373), then its subsequent action depends on whether or not the next ticket has been claimed. If the next ticket has already been claimed (line 375), then it attempts to release the holder of that ticket from its busy-waiting loop if needed (line 376). Further, if it is the holder of the next ticket and has already completed its critical section, then it attempts to advance the token (line 378). On the other hand, if the next ticket has not been claimed, then it attempts to cancel the ticket (line 383). If the ticket has been canceled (possibly by another process) (line 384), then it attempts to advance the token (line 385). In both cases, if the process succeeds in advancing the token using a CAS instruction, it recursively invokes the `MoveToken` procedure again. Otherwise, the process returns.

Theorem 8.1 asserts the correctness properties of Chan and Woelfel's RME algorithm.

Theorem 8.1 *Chan and Woelfel's RME algorithm satisfies ME, SF, BCSR, and 0-BR properties.*

Theorem 8.2 asserts the RME complexity behaviour of Chan and Woelfel's RME algorithm.

Theorem 8.2 *Chan and Woelfel's RME algorithm has worst-case RMR complexity of $O(F + 1)$ per passage and amortized-case RMR complexity of $O(1)$ per passage, where F denotes the total number of crashes that have occurred in the system.*

8.3   Discussion

The *worst-case* RMR complexity of a passage in Chan and Woelfel's algorithm is unbounded. This is because the ticket numbers generated by a process may be repeatedly canceled due to another process failing repeatedly. As such, this may cause the (former) process to execute the repeat-until loop at lines 352 to 357 an arbitrary number of times. Further, the effect of a failure may last for an arbitrarily long time. In particular, a failure may cause a ticket number issued much later in the future to be canceled.

Reference

1. David Yu Cheng Chan and Philipp Woelfel. Recoverable mutual exclusion with constant amortized RMR complexity from standard primitives. In *Proc. of the 39th ACM Symposium on Principles of Distributed Computing (PODC)*, New York, NY, USA, August 2020.

Abortable Recoverable Mutual Exclusion

9

Locks typically require that if a process wishes to enter the critical section, it must spin and wait if there is another process already in the critical section. Under the context of recoverable mutual exclusion, such a process eventually either enters its critical section or crashes, assuming that the algorithm is starvation-free. The abortable recoverable mutual exclusion problem introduces an additional option for such a process to abort its attempt to acquire the lock. With this option, a process may choose to delay its attempt (by aborting) to enter into critical section instead of spinning idly, in order to do some other work and possibly retry later to acquire the lock. The advantage of this option is that a process may unblock itself when other processes are holding the lock for an arbitrarily long period of time.

A process p may choose to abort its attempt at any time when trying to acquire the lock. In order to signal the abortable recoverable lock to abort the attempt, a special variable called $AbortSignal[p]$ is used, which denotes whether p has received an abort signal. To that end, once $AbortSignal[p]$ is $\texttt{true}$, process p must not spin idly and abort its attempt. We assume that every abortable recoverable lock is equipped with a distinct $AbortSignal$ array with one entry for each process. Process p could receive its abort signal either by an interrupt signal, by another thread of the same process, or even by another process. The variable $AbortSignal[p]$ is assumed to be reset at the end of each super-passage by the environment. The task of setting and resetting the variable $AbortSignal[p]$ is beyond the scope of this work.

9.1 Problem Definition

This section presents a formal definition of the abortable recoverable mutual exclusion problem, building on the foundations established in Chap. 4.

© The Author(s), under exclusive license to Springer Nature Switzerland AG 2023
S. Dhoked et al., *Recoverable Mutual Exclusion*, Synthesis Lectures on
Distributed Computing Theory, https://doi.org/10.1007/978-3-031-20002-1_9

9.1.1   Control Flow

Just like algorithm in Fig. 4.1 from Chap. 4, when a process leaves the NCS, it executes the Recover, Enter, CS and Exit sections in that order.[1] However, it may happen that, while executing the Recover or Enter section, a process receives the abort signal. In that case, the process can simply abort its attempt and does not need to execute the CS or Exit sections. Note that despite receiving an abort signal, a process may have acquired the lock already. Based on the timing of the abort signal and the interleaving of processes while accessing the lock, a process may either abort its attempt or enter into CS.

To be able to determine if an attempt has been aborted, the Enter section is augmented with a boolean return value. This value informs the invoking process whether or not the process has been granted entry into CS. If Enter returns true, the process then continues to execute the CS and Exit sections in that order. Otherwise, the process goes back to the NCS. Algorithm 1 shows the execution path for an abortable recoverable lock.

Algorithm 1: Execution path of a process participating in the abortable recoverable mutual exclusion. A process reverts back to the NCS after recovering from a crash failure.

---

loop forever
    Non-Critical Section (NCS)
    Recover( )
    if Enter( ) **then**
        Critical Section (CS)
        Exit( )

Note that since a process may jump from the entry section to the non-critical section, without executing critical or exit section, the traditional definition of a passage (begins at NCS and concludes either at completion of Exit or a crash) needs to be modified.

Definition 9.1 A *passage* is a sequence of steps taken by some process p_i that begins with Recover and ends when either (a) Enter returns false, or (b) Exit executes till completion, or (c) p_i crashes.

If a process crashes at any point during its execution, it recovers and begins its execution again from the NCS. The CSR (or BCSR) property ensures that if a process crashes in the CS, its subsequent attempts cannot abort and jump to the NCS until a failure-free passage

---

[1] Note that the algorithms presented in this chapter were originally designed for a different execution model. They have been modified to fit into the Golab-Ramaraju model [1, 2].

has been executed. Thus, the next time this process executes the `Enter` section, `Enter` must return `true`.

9.1.2   Correctness Properties

Any abortable recoverable mutual exclusion algorithm must satisfy the correctness properties mentioned in Sect. 4.3. Furthermore, the added option of abortability imposes certain additional constraints on the RME algorithm. Firstly, we require that no process should abort its attempt trivially if the *Abort Signal* variable is not `true`. Second, a process should abort within a "reasonable" time once the *Abort Signal* variable becomes `true`. Lastly, we need to modify the SF and k-FCFS properties to account for the case when a process p may abort its attempt during the `Enter` section and never go to the CS. To capture these constraints, we define the following correctness properties (based on Jayanti and Joshi's work [3, 4] and Katzan and Morrison's work [5]) that an abortable recoverable mutual exclusion algorithm must satisfy:

Non-Trivial Abort (NTA): For any history H, if an execution of `Enter` by a process p returns `false` at some point in H, then *Abort Signal*$[p]$ must be `true` at that point in H.

Bounded Abort (BA): For any history H, any execution of `Recover` or `Enter` by a process p completes in a bounded number of p's steps once *Abort Signal*$[p]$ is `true`.

Starvation Freedom (SF): For any infinite fair history H *with finitely many failures*, if a process p leaves the NCS in some step of H then eventually p leaves the `Enter` section.

Jayanti and Joshi Starvation-Freedom (JJ-SF): For any infinite fair history H *where each process fails finitely often in any super-passage*, if a process p leaves the NCS in some step of H then eventually p leaves the `Enter` section.

k-First Come First Served (k-FCFS): For any history H, suppose that process p_i begins its ℓ_ith passage and p_j begins its ℓ_jth passage. Suppose further that neither passage is k-failure-concurrent, neither process receives an abort signal during its respective passage and that process p_i completes the doorway in its ℓ_ith passage before process p_j begins the doorway in its ℓ_jth passage in H. Then p_j does not enter the CS in its ℓ_jth passage before p_i enters the CS in its ℓ_ith passage.

9.2   Jayanti and Joshi's Algorithm

Jayanti and Joshi's solution is based on the concept of f-arrays [6]. An f-array stores an array of variables $[x_1, x_2, \ldots, x_n]$, one entry for each process, and can atomically compute the value of some aggregate function $f(x_1, x_2, \ldots, x_n)$. In their solution to abortable recoverable mutual exclusion, Jayanti and Joshi use a min-array, where $f(x_1, x_2, \ldots, x_n) = \min\{x_1, x_2, \ldots, x_n\}$, to determine the order of entry into CS.

The min-array supports two operations, FindMin() and Update(val). The FindMin operation returns the minimum value of the array. The operation Update(val), when executed by process p_i, writes val to the entry associated with p_i in the min-array.

9.2.1 A Recoverable Min-Array

In order to instantly retrieve the minimum value at any time, the min-array is implemented using a binary tree, such that, the minimum value is always maintained at the root node. The binary tree contains n leaf nodes, $leaf[1]$, $leaf[2]$, $\ldots leaf[n]$. This implementation guarantees $O(1)$ RMRs for the FindMin operation and $O(\log n)$ RMRs for the Update operation. Figure 9.1 presents an example of storing the min-array as a binary tree. The pseudocode of the implementation is given in Fig. 9.2.

The purpose of the Update(val) operation is two-fold. Firstly, it is used by p_i to update its entry in the min-array by storing val in $leaf[i]$. Secondly, it is used to update the value of each node along the path from $leaf[i]$ to $root$, so that the minimum value of the array may later be computed easily. For each internal node, process p_i stores the minimum value present in the subtree rooted at that node. This ensures that the minimum value of the entire array is maintained at the $root$ of the tree. The FindMin operation simply returns this minimum value stored at the $root$.

A key technique used in the min-array is the "double refresh". Each process uses a double refresh during Update(val) on line 393 and line 394. This is required since the CAS instruction on line 399 may be unsuccessful during the first execution of Refresh due to a conflicting CAS instruction executed by another process. The conflicting CAS may not have examined the value written by the current process. However, this does not happen during the second execution of Refresh. If the CAS during the second execution of Refresh is

Fig. 9.1 Example of a min-array with $n = 6$

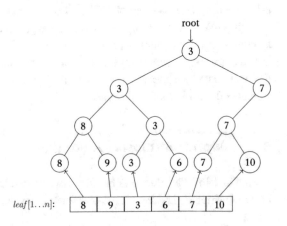

Procedure FindMin() for process p_i

---

388 return $root.val$

Procedure Update(val) for process p_i

---

389 $leaf[i].val := val$
390 $currentNode := leaf[i]$
391 **while** $currentNode \neq root$ **do**
392 | $currentNode := currentNode.parent$
393 | Refresh $(currentNode)$
394 | Refresh $(currentNode)$

Procedure Refresh($currentNode$) for process p_i

---

395 $currentval := currentNode.val$
396 $leftval := currentNode.left.val$
397 $rightval := currentNode.right.val$
398 $minval := \min(leftval, rightval)$
399 CAS $(\&currentNode.val, currentval, minval)$

Fig. 9.2 An implementation of recoverable min-array

also unsuccessful, it is assured that the latest conflicting CAS has now examined the value written by the current process and updated the minimum value to the node accordingly.

In the absence of failures, the implementation of min-array is linearizable and wait-free.

Observe that Update(val) operation is idempotent in the sense that executing it fully once is equivalent to executing it multiple times, possibly partially in some cases, with at least one instance executing to completion.

Theorem 9.1 *The worst-case RMR complexity of* FindMin *and* Update *is* $O(1)$ *and* $O(\log n)$, *respectively, in both CC and DSM models.*

9.2.2 Abortable RME

The primary idea behind this abortable recoverable mutual exclusion algorithm is as follows. The algorithm maintains an integer variable TOKEN that increases as processes attempt to enter the CS. Each process reads the current TOKEN value and stores it in a min-array called REGISTRY. The processes then enter the CS in an increasing order of their token value recorded in REGISTRY. When a process exits the CS, the next smallest token value in REGISTRY is read and the corresponding process is allowed to enter the CS. The processes use helping to determine the order of entry into CS. The helping is done by the Promote procedure. This helps to account for failures, as shown later (Fig. 9.3).

Procedure Abort() for process p_i

---

419 REGISTRY.Update $(\langle \infty, i \rangle)$
420 Promote(true)
421 **if** CSSTATUS $= \langle$true$, i \rangle$ **then return** false
422 Go$[i] := -1$
423 **return** true

Procedure Promote(flag) for process p_i

---

424 $\langle b, s \rangle :=$ CSSTATUS
425 $\mathit{peer} := s$
426 **if** $\neg b$ **then**
427 | $\langle \mathit{token}, \mathit{peer} \rangle :=$ FindMin()
428 | **if** $\mathit{token} = \infty \land \neg\mathit{flag}$ **then return**
429 | **if** $\mathit{token} = \infty$ **then** $\mathit{peer} := i$
430 | **if** $\neg$CAS$(\&$CSSTATUS$, \langle$false$, s \rangle, \langle$true$, \mathit{peer} \rangle)$ **then return**
431 $g :=$ Go$[\mathit{peer}]$
432 **if** $g \in \{-1, 0\}$ **then return**
433 **if** CSSTATUS $= \langle$true$, \mathit{peer} \rangle$ **then**
434 | CAS$(\&$Go$[\mathit{peer}], g, 0)$

Fig. 9.3 Jayanti and Joshi's abortable recoverable mutual exclusion algorithm. Presentation continued from Fig. 9.4

REGISTRY is a min-array that stores a pair of values, (i, token). The comparison ("less than" relation) between two values in REGISTRY is defined as follows: $\langle \mathit{token}, i \rangle < (\langle \mathit{token}', i' \rangle$ if $\mathit{token} < \mathit{token}'$ or $(\mathit{token} = \mathit{token}') \land (i < i')$. By default, REGISTRY stores $\langle \infty, i \rangle$ for every process p_i. The token value is also stored by process p_i in the shared variable Go$[i]$ which it later spins on (line 408). If Go$[i] = -1$, then the process is either in its recovery or non-critical section, or has aborted or exited its attempt and is about to enter its non-critical section. If Go$[i] > 0$, p_i then p_i is attempting to enter the CS. Process p_i is given access to the CS by setting Go$[i] = 0$.

Additionally, the algorithm uses shared variables CSSTATUS and SEQ, which operate in tandem. CSSTATUS holds a pair of values. If CSSTATUS $= \langle$true$, i \rangle$, then process p_i owns CS. On the other hand, if CSSTATUS $= \langle$false$,$ SEQ$\rangle$, then no process is in the CS. SEQ is an auxiliary integer variable that is incremented each time a process leaves the CS. It is used to avoid the ABA problem.

During the Enter section, process p_i reads the token value, increments the token value, and then writes the value read to Go$[i]$. Next, it updates REGISTRY, executes the Promote(false) procedure and spins till Go$[i]$ is reset to 0 or $\mathit{AbortSignal}[i]$ is set to true, whichever happens first. If the process stopped spinning as a result of Go$[i]$ being reset to 0 (line 409), then the execution of Enter returns true and the process owns the CS. On the other hand, if the process stopped spinning due to $\mathit{AbortSignal}[i]$ being true, it cannot simply return false. Since it may have been promoted to the CS by another process, it must first try to abort its attempt. If the abort is successful, then Enter returns false. Else, p_i owns the CS and Enter returns true (Fig. 9.4).

Shared variables:

- TOKEN: integer object, initially 1
- REGISTRY: min-array$[1\ldots n]$ of tuples, each tuple of the form $\langle$integer, integer$\rangle$, all elements initially $\langle \infty, i \rangle$
- Go: array $[1\ldots n]$ of integer objects, all elements initially -1
- CSSTATUS: tuple of $\langle$boolean, integer$\rangle$, initially $\langle$false, 1$\rangle$
- SEQ: integer object, initially 1

Private variables:

- $token$, s, $peer$, g: integer, uninitialized
- b: boolean, uninitialized

Procedure Recover() for process p_i

---
400   **if** Go$[i] \neq -1$ **then**
401 | Abort()

Procedure Enter() for process p_i

---
402   **if** CSSTATUS $= \langle$true, $i\rangle$ **then return** true
403    $token :=$ TOKEN
404   CAS(TOKEN, $token$, $token + 1$)
405   Go$[i] := token$
406   REGISTRY.Update($\langle token, i \rangle$)

407   Promote(false)

408   **await** Go$[i] = 0$ **or** $AbortSignal[i]$

409   **if** Go$[i] = 0$ **then**
410 | **return** true
411   **else**
412 | **return** $\neg$ Abort()

Procedure Exit() for process p_i

---
413   REGISTRY.Update($\langle \infty, i \rangle$)
414    $s :=$ SEQ
415   SEQ $:= s + 1$
416   CSSTATUS $:= \langle$false, $s + 1\rangle$
417   Promote(false)
418   Go$[i] := -1$

Fig. 9.4 Jayanti and Joshi's abortable recoverable mutual exclusion algorithm. Presentation continued in Fig. 9.3

When a process tries to abort its attempt, the abort may or may not be successful. In other words, it may be possible that an aborting process gets promoted to own the CS. In such a case, the abort is considered unsuccessful. If an abort by p_i is unsuccessful, p_i is guaranteed to own the CS. On the other hand, if p_i was able to abort while holding some token value $token_i$, then it is guaranteed that p_i does not use the same token value $token_i$ to enter into the CS in subsequent attempts.

The `Abort` procedure returns `true` if it is successful and `false` otherwise. In order to abort, process p_i first removes its entry from REGISTRY and then invokes the `Promote(true)` procedure. If CSSTATUS $= \langle true, i \rangle$ as a result of the `Promote` procedure, then the abort is unsuccessful and p_i now owns the CS. Otherwise, CSSTATUS would not get set to $\langle true, i \rangle$ hereafter (for the current attempt). In that case, p_i can clear the Go[i] variable and its abort would be considered successful.

As mentioned earlier, the objective of the `Promote(false)` procedure is to select the next process to enter into CS, if any. If CSSTATUS $= \langle true, j \rangle$, then p_j already owns the CS and the execution of `Promote` by process p_i signals p_j by setting Go[j] $= 0$ using a CAS instruction. Otherwise, if CSSTATUS $= \langle false, \_ \rangle$, then p_i looks for the next process p_k in REGISTRY that is attempting to enter into the CS and tries to set CSSTATUS to $\langle true, k \rangle$. If successful, p_k owns the CS and p_i signals p_k by setting Go[i] $= 0$.

The `Promote(true)` procedure is only executed from the `Abort` procedure. The only difference in this case occurs when the `FindMin` from the REGISTRY returns ∞. If so, there are no conflicting attempts to enter the CS. In that case, p_i tries to own the CS itself (lines 429 to 430) instead of aborting its attempt.

After p_i finishes its CS, it removes its entry from the REGISTRY, clears the CSSTATUS and promotes another process to enter into CS. Before, the value at CSSTATUS is cleared and set to $\langle false, SEQ \rangle$, p_i increments SEQ so that CSSTATUS is not erroneously set back to `true`, p_i by another process at line 430. Finally p_i promotes another process if any to enter into CS before updating Go[i] $= -1$ to enter the NCS.

The values of variables CSSTATUS and Go[i] are used by p_i to determine how to recover from a crash. In order to recover from failures, process p_i simply tries to abort its attempt, if any. If Go[i] $= -1$ after a failure, then process p_i crashed in the NCS and no recovery work is required. Otherwise, if after a failure, Go[i] > 0, then process p_i must have failed in the middle of an attempt. The techniques required to recover from such a failure can get involved. Instead, the process tries to abort its own attempt and if successful, retries to enter the CS with a new token value. If the abort is unsuccessful, one of the following cases must have happened: 1. Process p_i got promoted while it was trying to abort its attempt. 2. Process p_i had failed in CS earlier. In both cases, it is guaranteed that CSSTATUS $= \langle true, i \rangle$ and the process will be able to enter into CS in a constant number of its own steps. Additionally, the **BCSR** property is maintained.

The algorithm guarantees that a process would get a new token during each super-passage. This is crucial to ensure that a process is not spuriously released from spinning (line 408) by another process executing the CAS instruction on line 434 of the `Promote` procedure.

Additionally, this algorithm satisfies 0-FCFS. Once process p_i updates its entry in REGISTRY, its doorway is said to be complete (lines 402 to 406). After this point, any process p_j that wishes to enter into CS would get a larger token value from that of p_i. Thus its REGISTRY value would not be the minimum until p_i completes its CS or aborts.

Theorem 9.2 asserts the correctness properties of Jayanti and Joshi's algorithm.

Theorem 9.2 *Jayanti and Joshi's abortable RME algorithm satisfies ME, JJ-SF, BCSR, BE, BR, 0-FCFS and CS continuity properties. Further, it also satisfies BA and **NTA** properties.*

Theorem 9.3 asserts the complexity behaviour of Jayanti and Joshi's algorithm.

Theorem 9.3 *Jayanti and Joshi's abortable RME algorithm has worst-case RMR complexity of O (log n) per passage in the CC and DSM models.*

9.3   Katzan and Morrison's Algorithm

Katzan and Morrison's solution achieves $O\left(\log n/\log\log n\right)$ RMR complexity per passage. Their algorithm is built in two steps, similar to the algorithms described in Chap. 6. The first step is a Δ-ported abortable recoverable lock, followed by a tournament tree.

The Δ-ported lock in this algorithm only incurs $O(1)$ RMRs per passage, independent of failures. This is achieved with an elegant bit-masking trick with the help of the *fetch-and-add* (FAA) instruction, as shown later. The value of Δ is restricted by the size of the word on which FAA can atomically operate. Let W be the maximum word size. Then $\Delta \leq W$. Assuming a standard word size, $W = O(\log n)$, the overall RMR complexity of Katzan and Morrison's algorithm is bounded by $O\left(\log n/\log\log n\right)$.

9.3.1   A Δ-Ported Abortable Recoverable Lock

Consider a process p_i attempting to acquire the Δ-ported lock via the port numbered k. We will sometimes use q_k to refer to p_i in the rest of this section. The main idea of the RME algorithm is to use a W-bit mask represented by the shared variable ACTIVE. If the kth bit in ACTIVE is set, then p_i is in contention to acquire the lock using port k. As long as p_i has a pending request, the kth bit can only be modified by p_i. The FAA instruction is used in order to set the kth bit in ACTIVE from 0 to 1 as follows: FAA(ACTIVE, 2^k). Similarly, FAA(ACTIVE, -2^k) resets the kth bit from 1 to 0.

Note that the return value of FAA might be lost due to failures. However, this algorithm is designed in such a way that the return value of the FAA instruction is not required. Instead, q_k simply checks the corresponding kth bit in variable ACTIVE to determine whether its FAA was performed. Additionally, the kth bit in ACTIVE can be set (or reset) using the above technique only if it is not already set (or reset). Otherwise, the FAA would interfere with other bits and the algorithm would have undesired behavior. To check if the kth bit is set in ACTIVE, we use the bit-wise "AND" operator, denoted "&" in the pseudocode. If (ACTIVE & 2^k) = 0, then the k-bit is not set in ACTIVE.

The variable Go[k] is now a pointer that points to a boolean value, which is used by the corresponding process for spinning (line 447). Process q_k allocates a new boolean object on the heap, stores it in Go[k] and later retires[2] it during each super-passage. The variable STATUS[k], (STATUS[k] $\in$ {ENTER, CS, EXIT},default= ENTER) stores the current state of process q_k with respect to the lock. Finally, the variable CSSTATUS now holds a tuple of the form $\langle taken, owner, go \rangle$ where $taken = 1$ represents that the CS is owned by some process. If the CS is "taken", then $owner$ represents the owner of the CS and go points to its corresponding boolean spin variable (Figs. 9.5 and 9.6).

The Recover, Enter and Exit methods for this lock have been extended to accept a parameter k that represents the port number that a process is using in order to acquire the lock. During the Enter section, q_k first allocates a new boolean variable and writes its address to Go[k]. Then it announces its intention to enter into CS by performing FAA on ACTIVE (as described above), executes the Promote($\perp$) procedure and spins till *Go[k] is reset to 0 or $AbortSignal[i]$ is set to true, whichever happens first. If q_k stopped spinning as a result of *Go[i] being true (line 448), then the Enter returns true and q_k now owns the CS. On the other hand, if q_k stopped spinning due to $AbortSignal[i]$ being true, it invokes the Exit(k) procedure as a sub-routine in order to undo its effect on the lock and then returns false.

The objective of the Promote($\perp$) procedure is to "elevate" the next process to enter into CS, if possible. Process q_k can only enter the CS if CSSTATUS $= \langle 1, q_k, Go[k] \rangle$. If CSSTATUS $= \langle 1, owner, go \rangle$, then $owner$ already owns the CS and the execution of Promote by process q_k signals $owner$ to enter into CS by setting *go = true (line 474). Otherwise, if CSSTATUS $= \langle 0, owner, go \rangle$, then q_k looks for the next process q_j (line 469) in ACTIVE that is attempting to enter into CS and tries to set CSSTATUS to $\langle 0, j, Go[j] \rangle$ using the CAS instruction at line 471. If successful, q_j owns the CS and q_k signals q_j by setting go = true (line 474).

After q_k finishes its CS, it removes its entry from the variable ACTIVE by performing another FAA (line 456), clears the CSSTATUS (line 460) and then promotes the next process to enter into CS (line 461). Finally, it clears the pointer value stored in Go[k]. The Promote(k) procedure on line 457 is only present to address a process executing Exit(k) as a result of an abort. It does not modify the behavior of a process exiting the CS since the exiting process already owns the CS.

In order to abort the attempt, q_k executes the Exit(k) procedure as a sub-routine. The call to Promote(k) helps avoid a crucial deadlock. Consider the following scenario. Process q_k executed FAA as part of the Enter procedure. This causes the Promote procedure of another process q_j to select q_k to be promoted. However, before q_j could promote q_k, q_k observed $AbortSignal[i]$ = true and invoked Exit(k). Later, q_j promotes q_k to own the CS just before p_k invokes Promote($\perp$) as part of Exit(k). The execution of Promote($\perp$) would have no effect in this case and q_k would simply abort while holding the lock. Note

---

[2] Even though a **boolean** on the heap is retired, it may not be safe to free/reuse its memory. A separate memory reclamation algorithm is required to free/reuse such those locations.

Shared variables:

- ACTIVE: integer object of W bits, initially 0
- STATUS: array$[1\ldots\Delta]$ of integer objects, each element in the set {ENTER, CS, EXIT}, all elements initially ENTER
- Go: array $[1\ldots\Delta]$ of pointers to boolean object, all elements initially **null**
- CSSTATUS: 3-tuple of the form ⟨boolean, integer, pointer to boolean⟩, initially ⟨false, 0, **null**⟩

Private variables:

- *owner*, *varactive*: integer object, uninitialized
- *taken*: boolean object, uninitialized
- *go*: pointer to boolean object, uninitialized

Procedure Recover(k) for process p_i

435	**if** STATUS$[k]$ = EXIT **then**
436	$\quad$ Exit(k)

Procedure Enter(k) for process p_i

437	**if** STATUS$[k]$ = CS **then**
438	$\quad$ **return** true
439	**if** $AbortSignal[i]$ **then**
440	$\quad$ Exit(k)
441	$\quad$ **return** false
442	**if** Go$[k]$ = **null** **then**
443	$\quad$ Go$[k]$:= a new boolean object
444	**if** (ACTIVE & 2^k) = 0 **then**
445	$\quad$ FAA (&ACTIVE, 2^k)
446	Promote($\bot$)
447	**await** *Go$[k]$ **or** $AbortSignal[i]$
448	**if** *Go$[k]$ **then**
449	$\quad$ STATUS$[k]$:= CS
450	$\quad$ **return** true
451	**else**
452	$\quad$ Exit(k)
453	$\quad$ **return** false

Fig. 9.5 The Δ-ported abortable recoverable lock used in Katzan and Morrison's abortable RME algorithm. Presentation continued in Fig. 9.6

that the above deadlock scenario can only occur if the lock is not already taken. To avoid this scenario, q_k executes the Promote(k) procedure in an attempt to update the CSSTATUS if the CS is empty. Here, either some other process or q_k itself starts owning the CS. If q_k owns the CS, it can now release the CS by updating CSSTATUS. However, since CSSTATUS is now updated, any subsequent attempts to promote q_k would be unsuccessful during the CAS on line 471.

Procedure Exit(k) for process p_i

---

454 STATUS[k] := EXIT
455 **if** (ACTIVE & 2^k) $\neq$ 0 **then**
456 | FAA (&ACTIVE, -2^k)

457 Promote(k)

458 $\langle taken, owner, go \rangle$:= CSSTATUS
459 **if** $taken$ **and** $owner = k$ **then**
460 | CAS(&CSSTATUS, $\langle$true, $owner$, $go\rangle$, $\langle$false, $owner$, $go\rangle$)

461 Promote($\bot$)

462 **if** Go[k] $\neq$ **null then**
463 | Go[k] := **null**

464 STATUS[k] := ENTER

Procedure Promote(j) for process p_i

---

465 $\langle taken, owner, go \rangle$:= CSSTATUS
466 **if** $\neg taken$ **then**
467 | $active$:= ACTIVE
468 | **if** $active \neq 0$ **then**
469 | | j := next($owner$, $active$)
470 | **if** $j \neq \bot$ **then**
471 | | CAS(&CSSTATUS, $\langle$false, $owner$, $go\rangle$, $\langle$true, j, Go[j]$\rangle$)

472 $\langle taken, owner, go \rangle$:= CSSTATUS
473 **if** $taken$ **then**
474 | $*go$:= true

Fig. 9.6 The Δ-ported abortable recoverable lock used in Katzan and Morrison's abortable RME algorithm. Presentation continued from Fig. 9.5

In the event of recovery after a failure, process q_k would execute Recover and Enter again. The recovery behavior of q_k depends on the value of STATUS[k]. If

- STATUS[k] = CS: Process q_k owns the CS and should be able to enter into CS within a bounded number of steps. This is true since Recover would not do anything in this case and Enter would simply return true.
- STATUS[k] = ENTER: Either the process is in the non-critical or entry section. Recover does not need to do anything in this case since the Enter procedure is idempotent.
- STATUS[k] = EXIT: Process q_k crashed while aborting or exiting its attempt. Since the Exit procedure is also idempotent, it is sufficient to merely re-execute the Exit procedure.

The next($owner$, $active$) function at line 469 finds the next bit that is set in $active$, starting from the bit corresponding to $owner$ and searching in a circular manner when reaching the end of the word. Since next($owner$, $active$) is only invoked if $active \neq 0$, it is guaranteed that there is at least one bit set in $active$. Details of the implementation of this procedure have been skipped. However, note that this procedure would incur only a constant number of RMRs, since the variable ACTIVE is copied into local variable *active* before invoking next(*owner, active*).

Theorem 9.4 asserts the correctness properties of the Δ-ported n-process abortable RME algorithm.

Theorem 9.4 *The Δ-ported n-process RME algorithm described in Figs. 9.6 and 9.5 satisfies ME, JJ-SF, BCSR, BE and 0-BR properties. Furthermore, it also satisfies BA and **NTA** properties.*

Theorem 9.5 asserts the complexity behavior of the Δ-ported n-process abortable RME algorithm.

Theorem 9.5 *The Δ-ported n-process RME algorithm described in Figs. 9.6 and 9.5 has worst-case RMR complexity of $O\,(1)$ per passage in the CC and DSM models.*

9.3.2   Tournament Tree

The "main" abortable recoverable lock designed by Katzan and Morrison is based on the concept of a tournament tree also used in Chap. 6.

The main difference between the two versions of the tournament tree is that, along the path from *leaf* to *root*, an invocation of the Enter procedure by a process may return false if the process wishes to abort. In that case, the process has to reverse its direction and release all the locks it has acquired so far. Pseudocode for the tournament tree is presented in Fig. 9.7.

If p_i receives an abort signal, its *node*. Enter would return false (line 482). As a result, it invokes the Exit procedure as a subroutine. The Exit procedure would then release all the locks, starting from the last acquired node (*Current*[i]) until the corresponding leaf node.

Theorem 9.6 asserts the complexity behavior of Katzan and Morrison's abortable RME algorithm.

Theorem 9.6 *Katzan and Morrison's abortable RME algorithm satisfies ME, JJ-SF, BCSR, BE and BR properties. Furthermore, it also satisfies BA and **NTA** properties.*

Theorem 9.7 asserts the complexity behavior of Katzan and Morrison's abortable RME algorithm.

Theorem 9.7 *Katzan and Morrison's abortable RME algorithm has worst-case RMR complexity per passage of $O\,(\log n/\log\log n)$ in both CC and DSM models.*

Shared variables:

- a Δ-ary tree with $\Delta \geq 2$ that contains at least $\lceil \frac{n}{\Delta} \rceil$ leaf nodes and whose every node is a Δ-ported RME lock described in figs. 9.5 to 9.6, *root* denotes the root node of the tree
- $leaf[1 \ldots n]$: array of pointers to tree nodes, each element i points to the bottommost (first) tree lock to be acquired by process p_i

---

Procedure `Recover()` for process p_i

     `// no specific recovery action is required`

Procedure `Enter()` for process p_i

---

475   $Current[i] := leaf[i]$
476   **repeat**
477      $node := Current[i].parent$
478      $k := port(i, node)$
479      $node.\text{Recover}(k)$
480   **if** $node.\text{Enter}(k)$ **then**
481          $Current[i] := node$
482   **else**
483          $\text{Exit}()$
484   **return** false
485   **until** $node = root$
486   **return** true

Procedure `Exit()` for process p_i

---

487   $node := Current[i]$
488   **while** $node \neq leaf[i]$ **do**
489      $k := port(i, node)$
490      $node.\text{Exit}(k)$
491      $node := node.children[k]$
492      $Current[i] := node$

Fig. 9.7 The tournament tree used in Katzan and Morrison's abortable RME algorithm

References

1. Wojciech Golab and Aditya Ramaraju. Recoverable mutual exclusion. In *Proc. of the 35th ACM Symposium on Principles of Distributed Computing (PODC)*, pages 65–74, 2016.
2. Wojciech Golab and Aditya Ramaraju. Recoverable mutual exclusion. *Distributed Computing (DC)*, 32(6):535–564, 2019.
3. Prasad Jayanti and Anup Joshi. Recoverable mutual exclusion with abortability. In *Proc. of 7th International Conference on Networked Systems (NETYS)*, pages 217–232, 2019.
4. Prasad Jayanti and Anup Joshi. Recoverable mutual exclusion with abortability. arXiv:2012.03140v1, 2020.
5. Daniel Katzan and Adam Morrison. Recoverable, abortable, and adaptive mutual exclusion with sublogarithmic RMR complexity. In *Proc. of the 24th International Conference on Principles of Distributed Systems (OPODIS)*, pages 15:1–15:16, 2021.
6. Prasad Jayanti. F-arrays: Implementation and applications. In *Proc. of the 21st ACM Symposium on Principles of Distributed Computing (PODC)*, PODC '02, pages 270–279, New York, NY, USA, 2002. Association for Computing Machinery.

Tight Lower Bound

10

Chan and Woelfel [1] recently proved a lower bound of $\Omega(\log n/\log\log n)$ on the RMR complexity of an RME algorithm in the worst case for both CC and DSM models when a process can use read, write, fetch-and-increment (FAI), fetch-and-store (FAS) and compare-and-swap (CAS) instructions on registers and processes fail independently. The lower bound holds even when each process executes *at most one super-passage* and *fails at most once* during its super-passage.

The lower bound is tight and proves that the RME algorithms by Golab and Hendler [2] and Jayanti, Jayanti and Joshi [3], both of which have sub-logarithmic RMR complexity, are asymptotically optimal. The lower bound does not apply to the RME algorithm by Katzan and Morrion [4], which also uses fetch-and-add (FAA) instruction.

On a very high level, the construction follows the outline of Anderson and Kim's lower bound on the RMR complexity for non-recoverable ME algorithms [5], which only applies to read-write registers that also support CAS instructions.

Schedules: The lower bound proof uses the notion of a *schedule*. A schedule is a sequence of steps of processes. A step may be regular step or crash step. A regular step may be read, FAI, FAS or CAS instruction (a write instruction is treated as a special case of FAS instruction). The proof then *inductively* constructs *families* of schedules, one family at each inductive step, satisfying certain properties until some process has executed $\Omega(\log n/\log\log n)$ steps that are guaranteed to incur RMR.

Each schedule in a family satisfies certain desirable properties. First, a process in a schedule executes at most one super-passage and crashes at most once. Second, a schedule contains two types of processes, namely *finished* and *active*. A finished process has completed its super-passage and returned to its NCS. An active process is still in its recovery or entry section and has executed at least one step. Third, an active process is *invisible* to all other active processes as well as all finished processes of a schedule, which is achieved as follows. No "information flow" is allowed from an active process to another active process. This

© The Author(s), under exclusive license to Springer Nature Switzerland AG 2023
S. Dhoked et al., *Recoverable Mutual Exclusion*, Synthesis Lectures on
Distributed Computing Theory, https://doi.org/10.1007/978-3-031-20002-1_10

constraint is required so that any schedule in which a given process is active has a *finite* extension in which that process executing in isolation performs at least one RMR incurring step. Without this constraint, an active process may start busy-waiting for another active (since the former is aware of the latter) and, thus, may not perform an RMR incurring step until the other process has advanced. To that end, an active process cannot access a register that has been last modified (CC model) or is owned by another active process (DSM model). Further, limited "information flow" may occur from an active process to a finished process. However, the finished process either fails before it could save the "knowledge" in persistent memory, or is unable to tell whether the step was performed by an active process or another finished process. To that end, if a finished process accesses a register that has been last modified (CC model) or is owned (DSM model) by another active process, then the finished process crashes before it could act on the value it observed.

As mentioned earlier, at each inductive step, the construction generates a *family of schedules*. The schedules belonging to the *same* family are related to each other in the following way. First, all schedules in the same family have the same set of finished processes. Second, no active process can distinguish between two different schedules of the same family in which it has taken steps. This implies that, if any schedule of a family is extended by appending steps of a subset of its active processes, then the same extension is also valid for another schedule that contains that subset provided the following holds: no process accesses a register in the extension that has been last modified (CC model) or is owned (DSM model) by an active process outside the subset. Third, the family contains a unique schedule whose active processes are a *super-set* of active processes of other schedules in the family. We refer to this special schedule as the *primary* schedule; note that each inductive step has its own primary schedule. Fourth, any subset of active process of a schedule S in the family can be erased (all their steps removed) in the following sense: There is another schedule S' in the family such that the set of active processes of S' is given by the set of active processes of S minus the set of erased processes. This is tricky if an active process being erased has performed one or more FAI instructions earlier. Removing those steps will change the register values and may cause the subsequent read from the register by another process to observe a different value, thereby rendering the suffix invalid. To address this problem, the construction ensures that any step of an active process that involves the FAI instruction can be replaced with an equivalent step of some finished process that also performs the FAI instruction. Note that this property implies that the family is complete in the sense that it contains a schedule for every possible non-empty subset of active processes in the primary schedule. Finally, for the family of schedules constructed at inductive step i, every active process in every schedule has performed at least i steps that incur an RMR.

Initially, the family contains empty schedules for every possible subset of processes in the system. Assume that the primary schedule at inductive step i contains n_i active processes. The construction ensures that primary schedule at step $i + 1$ contains at least $n_i/\log^{O(1)} n$ processes. This implies that the induction can be applied until $i = \Theta(\log n/\log \log n)$, thereby proving the desired lower bound.

Inductive Construction: The inductive construction first appends steps to the primary schedule of the current family such that the next step of every active process in the primary schedule, if executed, will incur an RMR. The subsequent steps of the construction depends on the degree of contention involving these next steps.

Low contention scenario: In this scenario, the set of registers accessed by active processes in their next step (which will incur an RMR) is relatively large. Thus, on average, only a small number of active processes access the same register. The proof then constructs a *conflict graph* where an edge exists between two active processes if their next step involves the same register or it will allow one process to discover the other. It can be shown that this graph is relatively sparse and thus contains a relatively large independent set of size $\Omega\left(n_i/\log^{O(1)} n\right)$ by applying Turan's theorem. All other processes that are not part of the independent set are erased, and the resultant schedule serves as the primary schedule for the inductive step $i + 1$.

High contention scenario: In this scenario, the set of registers accessed by active processes in their next step (which will incur an RMR) is relatively small, and thus most of these registers are accessed by a relatively large number of active processes. This is where the proof diverges from Anderson and Kim's lower bound proof [5].

The construction first identifies the instruction that will be executed by most of the active processes in their next steps; recall that there are only four different types of instructions a process can execute. The active processes executing one of the remaining three instructions are erased. It then partitions the set of active processes into groups such that processes in each group accesses the same register and each group contains $\Theta(\log^{O(1)} n)$ processes. Active processes that could not be put in any group are erased. It can be proved that the number of groups that were created is sufficiently large. Next, for at least a constant fraction of the groups, three active processes are identified—two *alpha* processes and one *beta* process. The idea is to allow the two alpha processes to finish their super-passages and hide the effect of the RMR incurring step of the beta process using the RMR incurring steps of the two alpha processes of the same group. After performing their RMR incurring step, all alpha processes fail (and, thus will not remember the outcome of the step they performed) and then complete their super-passages after recovering. All active processes, except for these alpha and beta processes, are erased from the schedule. The alpha processes are then rolled forward and moved from active to finished processes. The resultant schedule in which only beta processes are the active processes serves as the primary schedule for the inductive step $i + 1$. It can be proved that the number of beta processes in the new primary schedule is $\Omega\left(n_i/\log^{O(1)} n\right)$.

References

1. David Yu Cheng Chan and Philipp Woelfel. A tight lower bound for the RMR complexity of recoverable mutual exclusion. In *Proc. of the 40th ACM Symposium on Principles of Distributed Computing (PODC)*, 2021.
2. Wojciech Golab and Danny Hendler. Recoverable mutual exclusion in sub-logarithmic time. In *Proc. of the 36th ACM Symposium on Principles of Distributed Computing (PODC)*, pages 211–220, 2017.

3. Prasad Jayanti, Siddhartha Jayanti, and Anup Joshi. A recoverable mutex algorithm with sublogarithmic RMR on both CC and DSM. In *Proc. of the 38th ACM Symposium on Principles of Distributed Computing (PODC)*, pages 177–186, 2019.
4. Daniel Katzan and Adam Morrison. Recoverable, abortable, and adaptive mutual exclusion with sublogarithmic RMR complexity. In *Proc. of the 24th International Conference on Principles of Distributed Systems (OPODIS)*, pages 15:1–15:16, 2021.
5. James H. Anderson and Yong-Jik Kim. An improved lower bound for the time complexity of mutual exclusion. *Distributed Computing (DC)*, 15(4):221–253, 2002.

System Wide Failures

11

Golab and Hendler showed in [1] that the RME problem is solvable in $O(1)$ RMRs per passage using widely-supported primitives in a model where failures are system-wide, meaning that all processes crash simultaneously. The algorithms presented in this work assume that processes receive helpful failure detection information from the system in the form of an *epoch number*, which is a positive integer that increases after each system-wide failure. For example, the epoch number can be the value of a counter that is incremented by the operating system after every failure, or some integer representation of the last system boot time.

With regard to the Golab and Ramaraju model [2], the main change is that the procedures `Recover`, `Enter`, and `Exit` receive the current epoch number as an additional parameter. All passages executed between successive system-wide failures receive the same epoch number, whereas passages made after a particular failure use a higher epoch number than passages made before this failure. Thus, epoch numbers increase monotonically, but need not reflect the exact number of failures that have occurred since the beginning of the execution. A practical algorithm that records epoch numbers in registers of finite size can only tolerate a finite (but very large) number of failures.

We present the algorithms introduced in [1]. First, we describe an RMR-efficient barrier synchronization primitive in Sect. 11.1. Then, we present an ME to RME transformation in Sect. 11.2. Finally, in Sect. 11.4 we present a transformation that adds the Failures-Robust Fairness (FRF) property described earlier in Chap. 4.

11.1 Barriers

The specification of the barrier is stated in Definition 11.1, which refers to a *barrier procedure* that is called by processes at the point of synchronization. Two variations of this barrier are presented, and one is used as a building block of the other. Both variations are based on the premise that some distinguished process called the *barrier leader* (*leader* for short) must

S. Dhoked et al., *Recoverable Mutual Exclusion*, Synthesis Lectures on
Distributed Computing Theory, https://doi.org/10.1007/978-3-031-20002-1_11

open the barrier before other processes, called non-leaders, are able to progress beyond the barrier. The barrier leader is fixed for the duration of a given epoch. The barrier can be reused in different epochs with different leaders.

Definition 11.1 A barrier procedure satisfies the following properties for any given epoch e:

(i) no call to the barrier procedure in epoch e terminates before the leader has begun its call;

(ii) in a fair history, the leader's call in epoch e eventually terminates (or a failure occurs); and

(iii) in a fair history, if the leader's call in epoch e terminates then every other call in epoch e eventually terminates (or a failure occurs).

11.1.1 Barrier Variation 1: Known Leader

In the first variation, presented in Figs. 11.1 and 11.2 the barrier procedure is called BarrierSub, and is invoked with two arguments: the epoch number *epoch*, and the process ID of the leader, which is decided using some external mechanism. The leader opens the barrier at line 495 by writing a global shared variable, and then uses a distributed signaling mechanism to wake up non-leader processes that may be waiting on the barrier. The signaling mechanism borrows ideas from [3], except that it uses Compare-And-Swap for handshaking among two processes as opposed to implementing leader election from reads and writes. This barrier variation is designed for and needed in the DSM model only, where it achieves $O(1)$ RMR complexity.

After writing R, the leader executes the helper procedure BSub-Leader in which it synchronizes with non-leader processes that may be accessing the barrier. This helper procedure begins with a handshaking mechanism at lines 500 to 505, which uses the array $C[lid][1...n]$ of CAS objects to settle a potential race condition between the leader and non-leaders executing line 502 and line 510. If the leader is the first to swap *epoch* into $C[lid][j]$ (i.e., wins the handshake) then p_j will progress through the barrier without blocking, otherwise p_j will wait for a signal. The for-loop at lines 500 to 505 generates a list of non-leaders who win the handshake (i.e., with whom the leader loses the handshake) and stores their IDs in consecutive elements of $L[lid][1...n]$ at line 503. The position of a process p_j in this array is recorded in $I[lid][j]$ at line 504. Finally, the leader signals the first process in this list at line 508, and sets off a chain reaction by which the k-th process in the list signals process number $k + 1$.

A non-leader process p_i calls BSub-NonLeader if $R < epoch$ at line 493, and executes the handshake at line 510. If p_i swaps *epoch* into $C[lid][i]$ before the leader (i.e., p_i wins the handshake) then p_i waits for a signal at line 509, and then signals the next process

Shared variables:

- R: read/write register, initially 0
- $C[1\ldots n][1\ldots n]$: array of readable CAS objects, elements $C[i][1\ldots n]$ local to process p_i, all elements initially 0
- $I[1\ldots n][1\ldots n]$: array of read/write registers, elements $I[i][1\ldots n]$ local to process p_i, all elements initially 0
- $L[1\ldots n][1\ldots n]$: array of read/write registers, elements $L[i][1\ldots n]$ local to process p_i, all elements initially 0
- $S[1\ldots n]$: array of read/write registers, element $S[i]i$ local to process p_i, all elements initially 0

Private variables:

- j, k, tmp: integer, uninitialized

Procedure BarrierSub($epoch, lid$) for process p_i

```
      // try fast path
493   if R = epoch then return
      // take slow path
494   if lid = i then
495   │   R := epoch
496   │   BSub-Leader(epoch)
497   else
498   │   BSub-NonLeader(epoch, lid)
```

Fig. 11.1 RMR-efficient barrier with known leader. Presentation continued in Fig. 11.2

in the list created by the leader at lines 513 to 513. Otherwise, if the leader already swapped $epoch$ into $C[lid][i]$ then p_i returns immediately since the barrier is open and p_i is not required to participate in the distributed signaling mechanism.

Theorem 11.1 asserts the correctness properties of the barrier.

Theorem 11.1 *BarrierSub satisfies its specification (Definition 11.1).*

11.1.2 Barrier Variation 2: Unknown Leader

The second variation of the barrier, presented in Figs. 11.3 and 11.4 is designed for both the CC and DSM models, where it also achieves $O(1)$ RMR complexity. The barrier procedure is called Barrier, and is invoked with two arguments: the epoch number $epoch$, and a boolean $isLeader$ indicating whether the caller is the barrier leader. Thus, each process knows whether it is the leader, but non-leader processes do not know the ID of the leader. The barrier is opened by the leader by writing the current epoch number to a shared variable R. The variable R persists across failures but does not need to be reset since the epoch number grows monotonically. Synchronization with non-leader processes is straightforward in the CC model, and is contained in procedure BarrierCC. The algorithm for the DSM model

---

Procedure BSub-Leader($epoch$) for process p_i

499 $k := 1$
500 **for** j in $1 \ldots n$ **do**
501 $\quad$ $tmp := C[i][j]$
502 $\quad$ **if** $tmp = epoch \vee \neg$CAS$(\&C[i][j], tmp, epoch)$ **then**
503 $\quad\quad$ $L[i][k] := j$
504 $\quad\quad$ $I[i][j] := k$
505 $\quad\quad$ $k := k + 1$

506 **if** $k > 1$ **then**
507 $\quad$ $tmp := L[i][1]$
508 $\quad$ $S[tmp] := epoch$

Procedure BSub-NonLeader($epoch, lid$) for process p_i

509 $tmp := C[lid][j]$
510 **if** $tmp < epoch \wedge$ CAS$(\&C[lid][j], tmp, epoch)$ **then**
511 $\quad$ **await** $S[i] = epoch$
512 $\quad$ $k := I[lid][i]$
513 $\quad$ **if** $k < N$ **then**
514 $\quad\quad$ $tmp := L[lid][k + 1]$
515 $\quad\quad$ **if** $tmp \neq 0$ **then**
516 $\quad\quad\quad$ $S[tmp] := epoch$

Fig. 11.2 RMR-efficient barrier with known leader. Presentation continued from Fig. 11.1

is contained in procedure `BarrierDSM`, and is much more involved. This procedure uses `BarrierSub` as a subroutine, which we refer to as the *secondary barrier*.

The implementations of `Barrier`, `BarrierCC`, and `BarrierDSM` are presented in Figs. 11.3 and 11.4. In the CC model, synchronization is achieved easily using a global spin variable R. The leader increases the value of R to the current epoch number at line 536, and the other processes wait for this state change at line 538. The execution path in the DSM model is more involved since a global spin variable cannot be used in an RMR-efficient manner. Synchronization is instead achieved using a combination of two mechanisms. First, the leader increases the value of the shared variable R to the current epoch number at line 546, which opens the barrier for any other process that reads R at line 539 after R is updated. A secondary mechanism is needed to handle the situation where the other process arrives at the barrier before the leader, and does not know who the leader is. This is dealt with by electing a secondary leader using the CAS object C at line 547 to 551. Once the barrier is opened by the barrier leader, the secondary leader is unblocked at line 549, and all processes meet at the secondary barrier `BarrierSub` at line 555.

The DSM execution path begins with resetting the CAS object C at lines 540 to 543. The algorithm cannot rely on the leader from the previous epoch to reset C at the end of

Shared variables:

- R: read/write register, initially 0
- C: readable/writable CAS object, initially $\langle \bot, 0 \rangle$
- $E[1 \ldots n][0 \ldots 1]$: array of read/write registers, elements $E[i][0 \ldots 1]$ local to process p_i, all elements initially 0
- $S[1 \ldots n]$: array of integer spin variables, element $S[i]$ local to process p_i, all elements initially 0

Private variables:

- $secldr, tag, ltag, e0, e1$: integer, uninitialized

Procedure Barrier$(epoch, isLeader)$ for process p_i

---

517   **if** CC model **then**
518   $\quad$ BarrierCC$(epoch, isLeader)$
519   **else if** DSM model **then**
520   $\quad$ BarrierDSM$(epoch, isLeader)$

Procedure GetTag$(epoch, j)$ for process p_i

---

521   $e0 := E[j][0]$
522   $e1 := E[j][1]$
523   **if** $e0 = epoch$ **then**
524   $\quad$ **return** 0
525   **else if** $e1 = epoch$ **then**
526   $\quad$ **return** 1
527   **else**
528   $\quad$ **if** $e0 > e1$ **then**
529   $\quad\quad$ **return** 1
530   $\quad$ **else**
531   $\quad\quad$ **return** 0

Procedure SetTag$(epoch)$ for process p_i

---

532   $tag := $ GetTag$(epoch, i)$
533   $E[i][tag] := epoch$
534   **return** tag

Fig. 11.3 RMR-efficient barrier with unknown leader—main barrier procedure and helper functions. Presentation continued in Fig. 11.4

BarrierDSM because a crash failure can prevent the leader from progressing that far.[1] Resetting C is difficult because one has to distinguish between the case when C holds the ID of the leader from a prior epoch (which must be reset), versus the case when C holds the ID of the current leader (which must not be reset). The algorithm makes the distinction by appending a binary tag to the process ID, which is why C holds an ordered pair of the from $\langle ID, tag \rangle$. The tag is toggled when process p_i reaches line 544, and is recorded implicitly

---

[1] The algorithm can be simplified under the assumption that a crashed process eventually recovers and reaches the barrier again, but in this section we present a more general solution that does not rely on this assumption.

Procedure BarrierCC(*epoch*, *isLeader*) **for process** p_i

---

535 **if** *isLeader* **then**
536   │ $R := epoch$
537 **else**
538   └ **await** $R = epoch$

Procedure BarrierDSM(*epoch*, *isLeader*) **for process** p_i

---

    // try fast path
539 **if** $R = epoch$ **then return**
    // slow path, reset CAS object first
540 $\langle secldr, ltag \rangle := C$
541 **if** $secldr \neq \bot$ **then**
542   │ **if** $ltag \neq$ GetTag(*epoch*, *secldr*) **then**
543   │   └ CAS(&C, $\langle secldr, ltag \rangle$, $\langle \bot, 0 \rangle$)

    // elect secondary leader
544 $tag :=$ SetTag(*epoch*)
545 **if** *isLeader* **then**
       // open the barrier
546   │ $R := epoch$
547   │ CAS(&C, $\langle \bot, 0 \rangle$, $\langle i, tag \rangle$)
548   │ $\langle secldr, ltag \rangle := C$
       // signal secondary leader
549   │ $S[secldr] := epoch$
550 **else**
551   │ CAS(&C, $\langle \bot, 0 \rangle$, $\langle i, tag \rangle$)
552   │ $\langle secldr, ltag \rangle := C$
553   │ **if** $secldr = i$ **then**
554   │   └ **await** $S[i] = epoch$

    // wait for secondary leader
555 BarrierSub(*epoch*, *secldr*)

Fig. 11.4 RMR-efficient barrier with unknown leader—procedures for CC and DSM models

by storing a pair of epoch numbers in the array $E[i][0\ldots1]$. The two array elements hold distinct values (except in the initial state), and the index of the higher value indicates the state of the tag after p_i's last call to SetTag. The value of the tag a process p_i will hold in the current epoch after calling SetTag is determined by calling GetTag at line 542. This procedure reads $E[i][0\ldots1]$ at lines 521 to 522, and searches for an array element that holds the current epoch number. If such an element exists, its index is the value of the tag for p_i in the current epoch (line 524 and line 526). If not, then the value of p_i's tag in the current epoch is different than in the last epoch where p_i toggled its tag, and is computed at lines 528 to 531.

To reset C, process p_i first checks whether C holds the ID of some process p_j and its tag (as opposed to $\bot$ and a tag) at lines 540 to 541. If so, and the tag stored in C does not match the tag computed for p_j for the current epoch (line 542), then p_i attempts to reset C using CAS at line 543. The tag prevents the ABA problem in case some other process resets C

and p_j then wins the secondary leader election at line 547 or line 551 in the current epoch after p_i reads C at line 540 and before p_i executes the CAS at line 543.

Theorem 11.2 asserts the correctness properties of the barrier.

Theorem 11.2 *Barrier satisfies its specification (Definition 11.1) and has worst-case RMR complexity of $O(1)$ in the CC and DSM models.*

11.2   Adding Recoverability

The barriers presented thus far can be used to implement several useful transformations. The first transformation, presented in Fig. 11.5, converts a conventional mutex (base mutex) to a recoverable mutex (target mutex) by resetting the base mutex after each failure. The main algorithmic idea is similar to [2], except that the overhead of resetting the base mutex is reduced greatly by taking advantage of the stronger failure model. Internally, the transformation uses a conventional base mutex $mtxB$, a CAS object C, and the RMR-efficient barrier described earlier in Sect. 11.1. The target entry and exit sections simply invoke their counterparts on the base mutex $mtxB$. The recovery section uses the shared variable C to determine who will be responsible for resetting $mtxB$, and the barrier to ensure that no process accesses $mtxB$ unsafely while it is being reset.

In steady-state failure-free operation, C holds the current epoch number, and so the body of the recovery section is bypassed due to the conditional statements at line 557 and line 564. Following the first failure, C holds a value ≥ 0 and less than the current epoch number. Under this condition, processes proceed to line 558 to elect a leader, and the latter process updates the value of C to a negative value $(-epoch)$, indicating that recovery is in progress. The leader then resets $mtxB$ at line 559, and updates C to a positive value $(epoch)$ at line 560, indicating that recovery is complete. The leader then enters the barrier at line 561. A process that lost the leader election at line 558 enters the barrier as a non-leader at line 563.

Some processes may begin executing the recovery section after the leader has already been chosen in the current epoch, and before the leader reaches line 560. Such processes read a negative value $(-epoch)$ from C at line 556, and then enter the barrier as a non-leader at line 565. If the recovery section in the previous epoch was interrupted by a failure, then a process may also read a negative value $> -epoch$ and < 0 from C at line 556, in which case it proceeds to line 558, similarly to the case when C holds a value ≥ 0 and $< epoch$. The leader election is then repeated for the current epoch.

Theorem 11.3 asserts the correctness properties of the transformation presented in Fig. 11.5.

The properties of the transformation are stated in Theorem 11.3.

Theorem 11.3 *The target mutex implemented using the transformation presented in Fig. 11.5 preserves the following correctness properties of the base mutex $mtxB$: ME, SF, TE*

and BE. Furthermore, it has worst-case RMR complexity of O *(* $f(mtxB)$ *), where* $f(mtxB)$
denotes the sum of the RMR complexity of $mtxB$ *and the RMR cost of resetting* $mtxB$ *at
line 559.*

The statement of Theorem 11.3 intentionally omits the bounded recovery property
because the recovery section presented in Fig. 11.5 may perform busy-waiting inside the
barrier operations. In particular, a process that enters the barrier as a non-leader may busy-
wait at lines line 563 or line 565. However, the recovery section is bounded in passages that
begin in a state where C holds the value of the current epoch. This includes all passages
occurring in failure-free histories, provided that the initial epoch number matches the initial
value of the shared variable C. It also includes passages that begin after some process has
completed lines 559 to 560 after the most recent failure.

Shared variables:
- $mtxB$: conventional mutex (base mutex)
- C: readable/writable CAS object, initially 0

Private variables:
- cur, ret: integers, uninitialized

Procedure Recover ($epoch$) for process p_i

556   $cur := C$
557   **if** $-epoch < cur < epoch$ **then**
558 | **if** CAS($\&C, cur, -epoch$) **then**
559 | | reset $mtxB$ to its initial state
560 | | $C := epoch$
561 | | Barrier($epoch$, true)
562 | **else**
563 | | Barrier($epoch$, false)
564   **else if** $cur = -epoch$ **then**
565 | Barrier ($epoch$, false)

Procedure Enter ($epoch$) for process p_i

566   $mtxB$.Enter()

Procedure Exit ($epoch$) for process p_i

567   $mtxB$.Exit()

Fig. 11.5 Transformation of conventional mutex to recoverable mutex

11.3 Adding Critical Section Reentry

The second transformation adds the CSR property to an existing RME algorithm (base mutex). Its pseudo-code is presented in Fig. 11.6. Critical section ownership is tracked using a pair of shared variables: $inCSpid$ stores the ID of the process currently in the CS, or the negation of this ID if that process is re-entering the CS following a failure; $inCSepoch$ stores the epoch number of the process who last entered the CS. A barrier, R, is used to synchronize the process who is re-entering the CS with the remaining processes. The barrier and the base mutex $mtxB$ work in tandem to ensure mutual exclusion.

A process p_i accessing the target mutex first reads $inCSepoch$ and $inCSpid$ in the recovery section at line 568 to determine whether it is recovering from a failure in (or

Shared variables:

- $mtxB$: base mutex, recoverable
- R: Barrier object (section 11.1.2)
- $inCSpid$: read/write register, initially $\perp$
- $inCSepoch$: read/write register, initially 0

Procedure Recover($epoch$) for process p_i

---

568 **if** $inCSepoch < epoch \wedge inCSpid = i$ **then**
          // CS re-entry after failure, proceed to entry section
569   **return**
570 **else if** $inCSpid \neq \perp$ **then**
571   **if** $inCSepoch < epoch$ **then**
572   R.Barrier($epoch$, false)

573 $mtxB$.Recover($epoch$)

Procedure Enter($epoch$) for process p_i

---

574 **if** $inCSepoch < epoch \wedge inCSpid = i$ **then**
          // CS re-entry after failure
575   **return**

576 $mtxB$.Enter($epoch$)
577 $inCSepoch := epoch$
578 $inCSpid := i$

Procedure Exit($epoch$) for process p_i

---

579 $inCSpid := \perp$
580 **if** $inCSepoch < epoch$ **then**
          // CS re-entry after failure
581      $inCSepoch := epoch$
582   R.Barrier($epoch$, true)
583      $mtxB$.Recover($epoch$)
584      $mtxB$.Enter($epoch$)

585 $mtxB$.Exit($epoch$)

Fig. 11.6 Transformation of recoverable mutex to CSR recoverable mutex

dangerously near) the CS. The order in which these two variables are read at line 568 does not matter. If $inCSepoch$ corresponds to a past epoch and p_i reads its own ID from $inCSpid$, then this indicates that p_i must re-enter the CS before any other process enters the CS. In that case, p_i leaves the recovery section at line 569 and continues in the entry section, where it proceeds to the CS after evaluating the condition at line 574. Note that p_i has not yet executed the recovery or entry section of the base mutex $mtxB$. In the exit section, p_i resets $inCSpid$ at line 579, then updates $inCSepoch$ to the latest epoch at line 581, opens the barrier R at line 582, and completes a super-passage through the base mutex $mtxB$ at lines 583 to 585. Updating $inCSepoch$ ensures that p_i does not incorrectly re-enter the CS again in the same epoch. The barrier R is used to barricade other processes inside the recovery section (see line 572) in order to allow p_i to re-enter the CS unimpeded, and must be opened to allow such processes to compete for the CS in the current epoch. Lines 583 to 584 are required only for correct recovery of $mtxB$, and can be omitted if the base mutex $mtxB$ satisfies the critical section continuity property.

A process p_i that does not read its own ID from $inCSpid$ at line 568 of the recovery section must check whether another process p_j may be recovering from a failure in the CS, in which case p_i must wait for p_j instead of competing for the base mutex. This case is identified at lines 570 to 571, which check whether $inCSpid \neq \perp$ and $inCSepoch < epoch$. These two conditions imply that p_j exists and must be given priority over p_i in this passage. In that case p_i waits for p_j at a barrier at line 572. As explained earlier, p_j eventually opens the barrier at line 582 of the exit section, at which point p_i is free to compete for the base mutex with other processes. Then, p_i completes the recovery section by executing the recovery section of $mtxB$ at line 572. In the entry section, p_i then executes the base entry section at line 576, ensures that the $inCSepoch$ variable stores the current epoch number at line 577, and then writes its ID to $inCSpid$ at line 578. In the exit section, p_i resets $inCSpid$ at line 579 and then completes the base exit section at line 585.

Theorem 11.4 asserts the correctness properties of the transformation. The proof is omitted due to lack of space.

The properties of the transformation are stated in Theorem 11.4.

Theorem 11.4 *The target mutex implemented using the transformation presented in Fig. 11.6 ensures CSR, and preserves the following correctness properties of the base mutex $mtxB$: ME and SF. Furthermore, it has the same RMR complexity asymptotically as $mtxB$.*

11.4   Adding Failure Robust Fairness

The FRF property can be added to a recoverable mutex using a transformation somewhat similar to the one presented in Sect. 11.3. The algorithm is presented in Fig. 11.7, and uses a barrier R in addition to a base mutex $mtxB$. The shared array $h[1\ldots n]$ is used to

Shared variables:

- $mtxB$: base mutex, recoverable
- R: Barrier object (section 11.1.2)
- $h[1...n]$: array of boolean objects, initially false
- $hIndex$: integer object, initially 1
- $hEpoch$: integer object, initially 0

Private variables:

- $waitOnBarrier$: boolean object

Procedure Recover($epoch$) for process p_i

586   $h[i] :=$ true
587   $mtxB.$Recover($epoch$)

Procedure Enter($epoch$) for process p_i

588   $mtxB.$Enter($epoch$)

Procedure Exit($epoch$) for process p_i

589   $h[i] :=$ false
590   $waitOnBarrier :=$ false
591   **if** $hEpoch < epoch$ **then**
592   **if** $hIndex = i$ **then**
593         $hEpoch := epoch$
594         $hIndex := (hIndex \mod n) + 1$
595         $R.$Barrier($epoch$, true)
596   **else if** $h[hIndex]$ **then**
597         $waitOnBarrier :=$ true
598   **else**
599         $hEpoch := epoch$
600         $hIndex := (hIndex \mod n) + 1$
601   $mtxB.$Exit($epoch$)
602   **if** $waitOnBarrier$ **then**
603      $R.$Barrier($epoch$, false)

Fig. 11.7 Transformation of recoverable mutex to FRF recoverable mutex

announce interest in the CS, and process p_i updates entry $h[i]$ at line 586, immediately upon invoking Recover. This flag is then cleared at line 589, immediately upon invoking Exit. The recovery and entry sections simply recover and enter the base mutex $mtxB$, and the exit section takes on the burden of ensuring FRF. Roughly speaking, after each crash failure, a privileged process p_i is chosen, and every other process p_j waits for p_i to pass through the CS before p_j finishes its passage. This ensures that p_j cannot complete super-passages infinitely often and bypass p_i perpetually. The process p_i is identified by the value of the shared variable $hIndex$, and the $hEpoch$ variable is used to ensure that $hIndex$ is incremented (modulo n) at most once per epoch.

The exit protocol has a special execution path at lines 592 to 600 that is protected by $mtxB$, and updates $hIndex$ during the initial passages of an epoch. This code also divides processes into three categories. The privileged process p_i identifies itself at line 592 based on the value of $hIndex$. It then updates $hEpoch$ and $hIndex$ at lines 593 to 594, and opens the barrier R as the leader for the current epoch at line 595. Any other process p_j detects whether the privileged process was in the middle of a passage at the time of the last failure by testing $h[hIndex]$ at line 596. If so, p_j sets a private variable at line 597 to remember that it must wait for p_i, then releases $mtxB$ at line 601, and finally waits for p_i at line 603 using the barrier R. Thus, non-privileged processes wait for p_i to complete the CS in the current epoch before finishing their own passage. If p_j does not detect a privileged process at line 596, it updates $hEpoch$ and $hIndex$ at lines 599 to 600 and releases $mtxB$ at line 601. The condition at line 591 is guaranteed to be false after the first complete execution of Exit in an epoch, in which case the barrier is bypassed.

One of the salient features of the exit protocol is that the shared variable $hIndex$ is incremented at most once per epoch, and exactly once if some process completes a super-passage in that epoch by executing Exit to completion. The first point follows from the use of $mtxB$ to protect lines 589 to 600, and the fact that $hEpoch$ is increased at line 593 or line 599 immediately before $hIndex$ is incremented at line 594 or line 600. The second point follows from the specific use of the barrier R, which ensures that no processes completes Exit in a given epoch until some process has incremented $hIndex$. This is because any process completing Exit for the first time in an epoch either increments $hIndex$ itself, or reaches line 597, and waits at line 603 for another process to perform the increment and then open the barrier at line 595.

The transformation achieves the FRF property with respect to the target mutex (but not the base mutex $mtxB$) by ensuring that a fast process p_i cannot complete super-passages infinitely often while a slow process p_j starves. The slow process sets $h[j] = \text{true}$ at the first line of Recover, and is able to enter the CS in the (eventual) absence of failures as long as the base mutex $mtxB$ provides starvation freedom. If failures occur repeatedly and yet other processes continue to complete super-passages infinitely often, then this implies that Exit is being executed to completion in infinitely many epochs. As explained earlier, the variable $hIndex$ is incremented at most once per epoch, and exactly once in any epoch where Exit is executed to completion, and so some process in our scenario eventually increments $hIndex$ from the value j to $j \mod N + 1$. This process must either be p_j itself at line 594, which implies that p_j eventually entered the CS, or another process at line 600. In the latter case, the condition tested at line 596 must have been false, which implies that $h[hIndex] = \text{false}$ held, and that once again that p_j eventually entered the CS.

The properties of the transformation are stated in Theorem 11.5.

Theorem 11.5 *The target mutex implemented using the transformation presented in Fig. 11.7 ensures FRF, and preserves the following correctness properties of the base mutex* $mtxB$: *ME, SF, CSR and BR. Furthermore, it has the same asymptotic worst-case RMR complexity as* $mtxB$.

References

1. Wojciech Golab and Danny Hendler. Recoverable mutual exclusion under system-wide failures. In *Proc. of the 37th ACM Symposium on Principles of Distributed Computing (PODC)*, pages 17–26, 2018.
2. Wojciech Golab and Aditya Ramaraju. Recoverable mutual exclusion. In *Proc. of the 35th ACM Symposium on Principles of Distributed Computing (PODC)*, pages 65–74, 2016.
3. Wojciech Golab, Danny Hendler, and Philipp Woelfel. An O(1) RMRs leader election algorithm. *SIAM Journal on Computing*, 39(7):2726–2760, 2010.

Discussion and Open Problems

12

Several RME algorithms described in this monograph have unbounded space complexity. They either use an infinite array [1] or require a process to allocate one or more objects on the heap at run time when executing a super-passage [2–5]. In the latter case, even though the process that allocated those objects may no longer need them after completing its super-passage, other processes may still hold references to those objects. If the process simply reclaims their memory after completion of its super-passage, then it may cause the RME algorithm to behave incorrectly. While Katzan and Morrison describe a memory reclamation approach for their abortable RME algorithm without adversely impacting its RMR complexity [5], it is not clear whether their approach can be extended to other RME algorithms [2–4]. An open problem is to design a recoverable memory reclamation algorithm for other RME algorithms that maintains all the correctness properties and complexity measures of the underlying RME algorithm.

The RME algorithms for system-wide failures described in Chap. 11, including the transformations to add CSR and FRF properties, assume the availability of a failure detector. This failure detector is modeled using an epoch counter, which is supplied as input to the `Recover`, `Enter` and `Exit` procedures. An open problem is whether it is possible to design an RME algorithm with $O(1)$ RMR complexity for system-wide failures, including transformations for CSR and FRF properties, that does not require a failure detector at all.

Other open problems include designing recoverable versions of other variants of the ME problem, including reader-writer mutual exclusion (RWME) and group mutual exclusion (GME). These variants allow processes to execute their critical sections concurrently under certain conditions.

© The Author(s), under exclusive license to Springer Nature Switzerland AG 2023
S. Dhoked et al., *Recoverable Mutual Exclusion*, Synthesis Lectures on
Distributed Computing Theory, https://doi.org/10.1007/978-3-031-20002-1_12

References

1. David Yu Cheng Chan and Philipp Woelfel. Recoverable mutual exclusion with constant amortized RMR complexity from standard primitives. In *Proc. of the 39th ACM Symposium on Principles of Distributed Computing (PODC)*, New York, NY, USA, August 2020.
2. Wojciech Golab and Danny Hendler. Recoverable mutual exclusion in sub-logarithmic time. In *Proc. of the 36th ACM Symposium on Principles of Distributed Computing (PODC)*, pages 211–220, 2017.
3. Prasad Jayanti, Siddhartha Jayanti, and Anup Joshi. A recoverable mutex algorithm with sub-logarithmic RMR on both CC and DSM. In *Proc. of the 38th ACM Symposium on Principles of Distributed Computing (PODC)*, pages 177–186, 2019.
4. Sahil Dhoked and Neeraj Mittal. An adaptive approach to recoverable mutual exclusion. In *Proc. of the 39th ACM Symposium on Principles of Distributed Computing (PODC)*, PODC '20, pages 1–10, New York, NY, USA, 2020. Association for Computing Machinery.
5. Daniel Katzan and Adam Morrison. Recoverable, abortable, and adaptive mutual exclusion with sublogarithmic RMR complexity. In *Proc. of the 24th International Conference on Principles of Distributed Systems (OPODIS)*, pages 15:1–15:16, 2021.

Printed in the United States
by Baker & Taylor Publisher Services